T0177217

IEE HISTORY OF TECHNOLOGY SERIES 30

Series Editors: Dr B. Bowers
          Dr C. Hempstead

RADIO MAN
The remarkable rise and fall of C. O. Stanley

Other volumes in this series:

RADIO MAN
The remarkable rise and fall of C. O. Stanley

Mark Frankland

Consultant
Gordon Bussey

The Institution of Electrical Engineers

Published by: The Institution of Electrical Engineers, London,
United Kingdom

© 2002: Stanley Foundation Ltd

British Library Cataloguing in Publication Data

Frankland, Mark, 1934-
Radio man: the remarkable rise and fall of C. O. Stanley. –
(IEE history of technology series; no. 30)
1. Stanley, Charles O. 2. Radio – Great Britain – History
I. Title II. Institution of Electrical Engineers III. Stanley Foundation
384. 5' 092

ISBN 0 85296 203 7

Typeset by Newgen Imaging Pte, India
Printed in England by Polestar Wheatons, Exeter, Devon

C.O. Stanley, c.1929

Contents

Acknowledgments

Two people made a quite unusual contribution to the writing of this book: Nicholas Stanley and Gordon Bussey.

Nicholas Stanley, backed by the Stanley family, collected most of the documents on which the biography of his grandfather is based. More than ten years ago, when almost all trace of C.O. Stanley seemed to have disappeared, Nicholas Stanley began the search for letters and documents that might throw light on his life. Given the disaster that overtook Pye in 1966 it was not surprising that the Stanley family were disinclined to keep records. Philips' takeover of Pye, and the latter's eventual disappearance even as a trade name, had apparently wiped out any institutional record of C.O. Stanley's successes as well as his failures.

There had been an earlier attempt to tell the story of his life. His son John Stanley asked Michael Bell, who worked for a time at Pye, to attempt a biography. However, Bell's work was hampered by the lack of documentary evidence, and the project lapsed with John Stanley's death.

By the time Nicholas Stanley told me of his plan for a new biography he had been able to put together a large collection of C.O. Stanley's personal and business papers, as well as many other documents on various aspects of his career. Equally valuable as a source were the interviews, conducted mainly by Nicholas himself, with several dozen of C.O.'s relatives and friends, colleagues and employees. Together with the collection of documents they have made it possible to put together a remarkably full account of a life that not long ago seemed forgotten beyond recovery.

The Stanley family made it plain that they wanted this biography to include all relevant material, regardless of how it reflected on C.O. Stanley; all judgments in the book are mine alone.

Gordon Bussey, the second indispensable contributor, is one of the leading experts on the history of radio. His extensive knowledge of the technology and development of the radio industry, and his familiarity with Pye receivers in particular, made him an invaluable adviser at every stage of the book's progress. He saved the author from many mistakes, and his suggestions on technical and radio industry matters have enriched the text.

He also applied his characteristic energy to the organisation of the large photo archive covering both Pye and C.O. Stanley's life. The quality of the book's illustrations is evidence of his skill. He has also written the captions.

Three other people also made essential contributions. The experienced archivist and researcher Pat Spencer brought order to the mass of paper assembled by Nicholas. She also retrieved from the Public Records Office documents illustrating C.O. Stanley's often stormy relationship with Whitehall and the armed services. (These documents are listed under Source materials.)

Stephen Hall applied his professional skills and first-hand knowledge of Pye to the study of the company's financial statistics, bringing them to ordered clarity. Both he and Gordon Bussey read the final text, though any errors that remain are mine.

I am also grateful to Michael Bell, whose unpublished work contains a great deal of invaluable information, not least insights into C.O. Stanley offered by family members, friends and colleagues who have since died.

The Stanley family would like to thank all who contributed to this project by giving interviews, and providing information and assistance. Their names are listed under Source materials.

Sir John and Lady Keane are warmly thanked for their hospitality in Cappoquin; Gordon Bussey received invaluable help from Claire Blackman, Rod Burman, Colin MacGregor, Guy Peskett, Nigel Rigler, Ian Robertson, Bob Smallbone, Gerald Wells, Louise Weymouth, Roger Woods and photographer Richard Williams.

Conversion table

Imperial unit	Metric unit
guinea	£1.05
shilling (s)	5 p
penny (d)	5/12 p
yard	0.9144 m
foot (ft)	0.3048 m
inch (in)	25.4 mm
pound (lb)	0.4536 kg

The execution

On 17 November 1966 a thousand people gathered in Holborn for the City of London's equivalent of an execution. No blood was to be shed, but the result would be scarcely less lethal: few of those who crowded into the Connaught Rooms that morning can have doubted they were about to witness the destruction of a man's reputation.

The occasion was as unusual as it was unpleasant. Several of the executioners, greeted by a mix of boos and cheers as they walked into the room, were old colleagues of the victim. As fellow directors they had together run the same company for 20 years, taken part in the same board meetings, and studied and approved the same production figures and accounts. Yet now the company they had jointly managed was in trouble it seemed they were innocent of all fault and only he, the planned victim, was guilty. The latter, appearing on the heels of his former colleagues and also to catcalls and cheering, though from different parts of the room, was tall and slim. A shock of fair hair, heavy spectacles and round face gave him the appearance of an earnest schoolboy. A less likely looking offender was hard to imagine.

The oddest thing about the occasion, though, was that it was Waterloo without Napoleon. The chief cause of the drama, and its earliest casualty, was nowhere to be seen. C.O. Stanley, creator of the Pye electronics empire and father of John Stanley, the man over whose neck the axe was poised in the Connaught Rooms, chose to remain a couple of miles away at his house in Lowndes Place. When the storm broke over Pye that spring Stanley had bowed out with scarcely a word of public explanation, resigning as managing director and chairman and accepting the face-saving post of company president. Since nothing important had ever happened at Pye without C.O. Stanley's approval the humiliation of his son was at least as much a judgment on the father.

Everyone supposed he wanted the son who had spent all his working life at Pye to succeed him. John Stanley's promotion five years earlier to deputy managing director had been a step in that direction, although anyone who understood anything about Pye knew it meant no slackening of the old man's control. If the son had indeed made mistakes it was obvious they had to be the father's too. Yet the father stayed away. It was a 'sacrifice' for him not to be there, he told a reporter, but he had to 'stand aside and avoid adding anything, which might only do more harm to the company'. The *Cambridge News*, Pye's local newspaper, quoted him as saying, 'I shall not be defending my son because he is quite capable of defending himself. But I won't be saying anything against him.'

With this less than energetic support from his father John Stanley's chance of escaping that day with his reputation intact never looked good. There was not even much hope to be taken from the last-minute intervention of Lord Goodman, solicitor to prime minister Harold Wilson and credited by some with near-miraculous powers as a mediator. The famous lawyer had volunteered to represent a group of shareholders opposing Pye's new chairman and the directors who had defected from the Stanley camp. In the event all Goodman felt able to do was demand that Pye appoint an independent director to its board. Of John Stanley he said not a word.

There were protests from the hall at the length of the lawyer's intervention, and the dignity of the proceedings was not helped by the poor performance of the amplifying equipment that Pye's own engineers had installed. There were many questions from the floor, some of them hostile to the team that had ousted the Stanleys, but the press, on the look-out for more evidence of the company's difficulties, gave prominence to a radio and TV dealer who said he had travelled 500 miles to London in order to deliver the message that the trade thought the once great firm of Pye had become 'a giggle'.

When John Stanley was at last allowed to defend himself, the *Times'* correspondent thought he made a fine speech, even if he did lapse into the language of student debates when he called his former colleagues on the Pye board 'nits'. He ended with a plea. 'Today is the day of the block for me. My life is in your hands.' He was quite right. The Stanley family's holding in Pye had dwindled to a small fraction of the total shares. C.O. Stanley might have run the company as though he still owned it, but the banks and institutions were its real masters, and having seen Pye shares collapse from the year's high of 19s 4d to little more than 6s it was not surprising they wanted the family out. When John Stanley sat down, the resolution to oppose his re-election as a director was carried by a margin of well over three to one. The *Mirror*, in a two-page spread headlined 'JOHN STANLEY SACKED IN THE PYE SHOWDOWN', printed a photograph of him striding out of the Connaught Rooms with a look of defiance.

The 'showdown' was more than a bitter end for the Stanleys. It was shocking to the wider public for whom the story of Pye and C.O. Stanley mirrored so much of the story of 20th century Britain. Pye could have been taken as a synonym for modernity from the moment in 1929 when C.O. Stanley, then a young advertising man, bought the small radio manufacturing business from the Cambridge scientific instrument maker W.G. Pye. Radio was a glamorous growth industry that was already revolutionising the way people led their lives and, less obviously, preparing the ground for the diverse electronics world to come. A sign of that future came with the outbreak of World War II when Stanley, already an extraordinarily self-confident leader, made sure Pye designed and built the radios needed for modern military tactics, and switched his newly formed team of television engineers to work on the land-, air- and sea-borne radar on which Britain's survival depended.

The names of C.O. Stanley and Pye were associated with some of the most dramatic applications of modern electronics in the post-war years. Pye pioneered radio telephones, equipping Britain's first fleet of radio cabs and bringing mobile phones to the police. Sir John Hunt took Pye radios with him to Mount Everest (though not quite to the summit). By the time of the crisis in 1966 Pye was the world's leading exporter of mobile radio phones. The company led the way in instrument landing systems for airports. It produced Britain's first transistor radio. In the 1950s it exported television cameras round the world, even selling them to the United States. A Pye camera was used to make the first picture-guided deep-sea recovery when a Comet, the pride of Britain's civil aviation in the 1960s, crashed in the Mediterranean.

By 1966 C.O. Stanley had built Pye into an international enterprise employing 30 000 workers in Britain and abroad. How could such a big and famous company run so spectacularly on the rocks? How could a man like C.O. Stanley, so cocky, so sure of himself, have failed to prevent the disaster?

Stanley was a public figure in his own right. Thanks to him, the *Observer* newspaper had written, Pye became 'the most newsworthy firm in a newsworthy industry'. It was certainly the most political – Stanley saw to that. His business interests led him into conflict with successive governments and agencies of the state. He played a leading part in industry's counter-attack against the first post-war Labour government, and was one of a handful of businessmen who turned a moribund Institute of Directors into the leading champion of free enterprise against what, for a time, looked like an unstoppable wave of nationalisation. Stanley was just as ready to turn against a Conservative government that dared support monopoly against free enterprise and the interests of C.O. Stanley. The Post Office, with its control of the airwaves needed by television and the growing radio telephone business, was the target of many of his attacks. The *Observer*, a liberal newspaper and not a natural fan, named him 'Businessman of the Year' when he broke the ring of telephone equipment manufacturers whose cosy relationship with the Post Office had left the country with a poor and expensive telephone system.

Stanley's most famous victory over monopoly had come when he helped finance, plan and energise the battle for commercial television. This pitted him against what seemed the majority of the British establishment. Stanley's grudge against the BBC was simple. Its television programmes were neither good nor popular enough to make people queue up to buy his TV sets. He happily took on a large part of both the Conservative and Labour parties as well as leaders of the churches, the universities and the Arts, and almost everyone else, who 50 years ago thought themselves responsible for the moral and intellectual well-being of the nation. Television made Stanley a star, as famous for what he did outside his factories as in them. By 1955 he had turned Pye into the only British television manufacturer involved in the entire television chain, from the making of programmes (Stanley was a founder of Associated Television) through the provision of cameras and studio equipment to transmitters and finally the sets on which the public watched what he called the new 'competitive' channel. He championed – often against the wishes of colleagues in his industry – each new development in television, from the opening of a third channel to the adoption of the modern 625 line picture system. In 1953 Pye had transmitted live colour pictures of the Queen's Coronation to the Great Ormond Street Children's Hospital. It was a typical piece of Stanley showmanship, and he remained to the end the industry's chief prophet of the long-delayed colour revolution.

Television made both boss and company rich. It provided Pye's greatest profits, above all in the ten years after the war when it manufactured the highest quality television sets on the market. And yet, if the company's new management were to be believed, television brought C.O. Stanley low. They published figures showing losses, chiefly on the television side, totalling millions of pounds. The bulk of this was the result of overproduction of television sets that cost too much and were of variable quality. A late attempt to break into the important TV rental market made a bad situation worse.

What went wrong? Unlike other famous names in the industry such as Mullard and E.K. Cole (the creator of Ekco, swallowed up by Pye), Stanley had no training in electronics. His element was the market: understanding it, exploiting it and anticipating it. When the *Economist* pronounced Pye's troubles to be the result of 'salesmanship unrestrained by cautious accountancy', anyone who knew anything about Pye recognised it as an indictment of C.O. Stanley himself. But while most of the many newspaper articles that analysed the company's fall from grace concentrated on Pye's inadequate financial procedures, no one explained how C.O. Stanley had come to lose his magic salesman's touch. If he was such 'a master of the art of creating demand for the products which his firm produces' (the opinion of the *New Scientist*), why was Pye now crippled by the burden of television sets no one would buy? And while it was plain that accountants had, to put it mildly, never counted for much in Pye, how did a company run in this way avoid trouble and even greatly flourish for over 30 years?

And there was another puzzle. All sorts of unpleasant insinuations were made against C.O. Stanley and his son, among them that Pye and its associated companies had been used to give employment to their family. Stanley's elder brother worked for Pye and so, at one time or another, had two of his sisters. The younger generation of the family also found employment there. Stanley did not deny it. His relatives, he told journalists, had often been underpaid (no evidence was produced to challenge this) and he 'believed in families'. How could this man who 'believed in families' leave his only son to face, on his own, the consequences of Pye's near collapse?

The more one studied the reports of the events of 1966 the more obvious it seemed that the explanation was less likely to be found in the accountants' reports and balance sheets than in the years that went before. That was where the story had to be buried; the end already being written in the beginning, final failure the undeniable child of earlier success. C.O. Stanley was indeed a 'family man', and it was his family that held the first clues to his remarkable rise and fall.

Chapter 1

An Irish family

Where do you begin the story of a life when that life is as out of the ordinary as Charles Orr Stanley's? Much that happened before he was born affected him greatly, and there are episodes in his earliest years that reveal as much about the trajectory of his later life as a mere date and place of birth, in his case 24 April 1899, in the little town of Cappoquin, County Waterford, near the southern edge of Ireland. So let the story begin one summer a year or so before the outbreak of the Great War, with a horse and cart loaded down with two army bell tents, a small rowing boat, pots and pans and all sizes of boxes. Five children, one boy and four girls, were on top of the cart as it set off from Cappoquin along the road leading east towards Dungarvan on the coast. Before this road crosses the Colligan River that feeds into Dungarvan Bay there is a farmhouse, and here the cart turned right to follow a track along the bay's south side. When the track came to an end at the shore of the Atlantic ocean the children climbed off the cart, but the horse turned left and continued on its own, so that a stranger to the place might have supposed it was about to walk into the water. In fact, horse and cart were proceeding along a narrow spit of sand that projected across the bay almost to the quayside of Dungarvan. This isthmus, called the Cunnigar, in places only a few yards wide, stretches for a mile and a half with no natural cover, unless patches of rough coastal grass and the black stumps of an old breakwater can be called cover.

A boy, perhaps ten years old, was waiting at the end of the spit where it was widest. The children who rode on the cart were his brother and sisters. They had climbed off for fear of loading it too heavily and they watched anxiously as the horse, which knew the way, made its slow progress along the plank of sand where the isthmus had been almost demolished by the attacks of wind and tide. This precarious path disappeared altogether at high tide and the receding water left treacherously wet sand to trap the cart. When Charlie, the boy waiting

Figure 1.1   C.O. Stanley with family and friends at Cunnigar, c.1909

Figure 1.2   C.O. Stanley with his sisters, c.1913

on his own at the end of the spit, saw the cart falter and sink a little, he muttered a prayer. The dray horse paused to gather its strength, and pulled free. The Stanley children's summer holiday had begun.

For the whole month of August they lived in the old army tents pitched on the sparse sands of the Cunnigar. In some years it rained for almost the entire month they spent there, and even when the weather was good it was scarcely a sweet place. But it was exciting; the children almost lived in the sea, and their days were dominated by its sound and smell and by the cry of birds in an open sky. It was also dangerous, as they discovered the summer they were told to look after a young cousin none of them much cared for. One day this boy pushed their dinghy off the beach, perhaps out of spite and in the hope it would be lost. Eddie, the elder Stanley boy, spotted the boat drifting out to sea and told Charlie, who was still in his bathing costume, to go into the water and bring it back. He ran into the sea, which was soon up to his waist. Reaching the boat just as a strong wind pushed it into the deep water of the main channel of the bay, he grabbed the gunwale and hung on. Too small to pull himself into the boat and yet too stubborn to let go, he might have drowned if his elder brother, a strong swimmer, had not managed to reach him and push him into the boat.

Eddie was barely 14 when they began these holidays, but apart from Sunday inspections by their parents who drove out in a pony and trap with fresh supplies and clean clothes, and an occasional visit from a local boatman who was asked to keep a distant eye on them, the children lived on the Cunnigar quite on their own. They cooked their own food and found much of it, too, rowing out to sea to set lobster pots or go fishing before dawn, when they usually came back with a catch of mackerel. They swam, played a simple sort of tennis on the sand, and with a lug sail rigged in the boat sailed the short distance across to Dungarvan itself to buy milk or visit the town's new attraction, the cinema.

The inventor of this daring form of holiday was the children's father, John Stanley, who owned the general stores in Cappoquin's Main Street. He was a restless, energetic man (Cappoquin's constable once arrested him for pedalling too fast on his penny farthing) and self-confident enough to like laying down the law. He was also a man on the up and up in a world full of obstacles to someone such as he. The Stanleys had settled in Ireland some time during the English colonisation. They were Protestants, members of the Church of Ireland, and this marked them

Figure 1.3   Main Street, Cappoquin, c.1900, where C.O. Stanley's father owned the General Merchants store

out from the Catholics who were their neighbours and who stood most to gain from the growing tide of national feeling that swept Ireland towards the turn of the century. Being Church of Ireland, though, was not enough to give them entry into Ireland's old ruling class, the Protestant Ascendancy of Anglo-Irish gentry who had supported centuries of British rule. Cappoquin lies between two great castles on the Blackwater River, Dromana and Lismore, the latter still belonging to the Dukes of Devonshire. In Cappoquin Ireland's ruling class was represented by the Keanes, who had large estates stretching over the Knockmealdown Mountains that the Stanley children could see from the Cunnigar, and a grand, grey Georgian house built on the site of an old FitzGerald castle. On a rise of land above the Blackwater, Cappoquin House looked across the river, but because the land fell away beneath its terraces it saw nothing of the town, whose houses gathered below like schoolchildren, and only a glimpse of the towers of the town's two churches, one Catholic and one Church of Ireland.

John Stanley did the only thing a man such as he could do to better himself: he tried to make money. It was not easy. Irish farming was backward, in some ways archaic. Industrial expansion was concentrated round Belfast in the north while even some of the most important towns near Cappoquin, such as Cork, were shrinking. But bettering oneself was a Stanley family tradition. John's grandfather, a shoemaker, had been ambitious for his children and taught his sons it was a mark of

Figure 1.4   River Blackwater, Cappoquin, c.1900

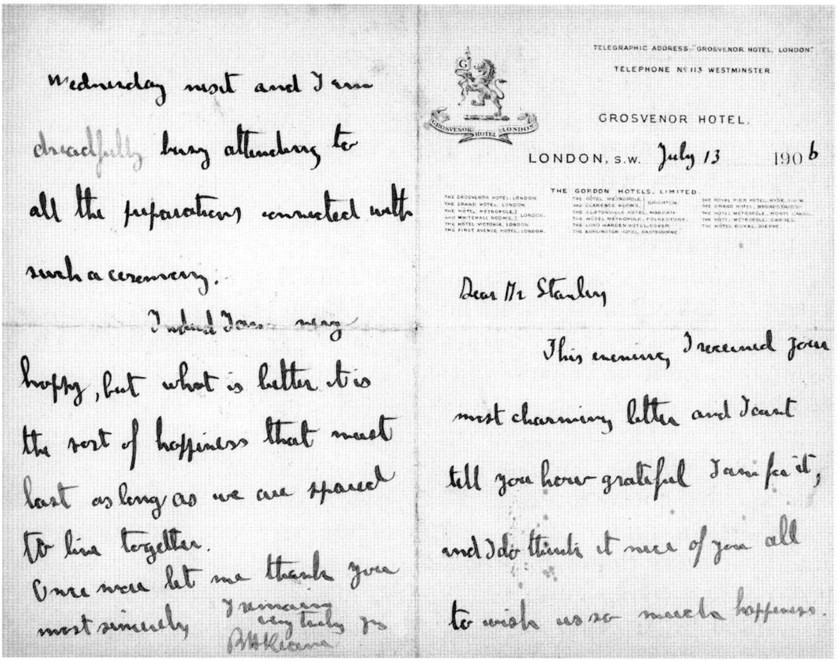

Figure 1.5   Letter to C.O. Stanley's father, John, from a member of the Keane family, 1906

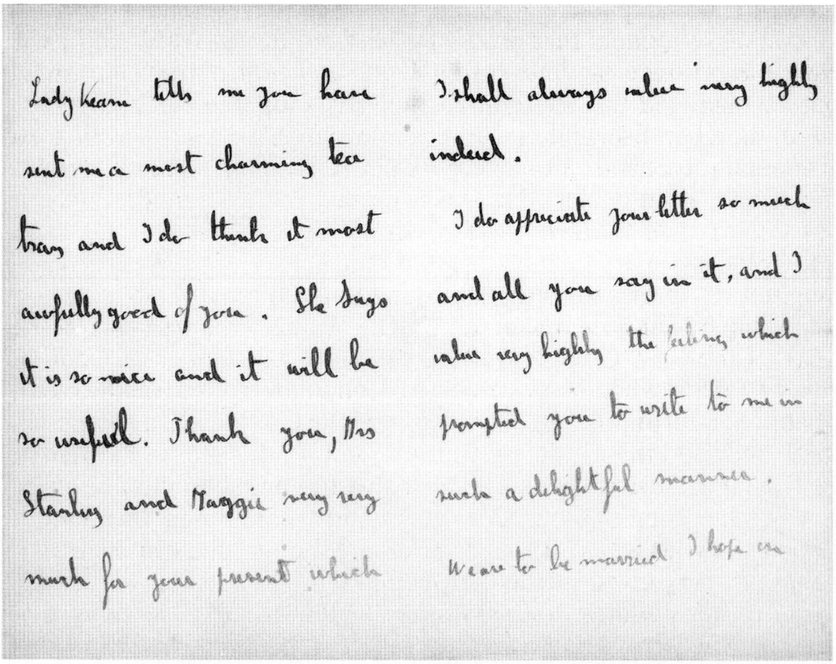

failure to end up as a soldier or a servant. One of them, William, who became a constable in charge of native police in India, admitted to his brother James that their father's words had 'struck to the heart' and given him 'a bad disposition which is I cannot bear to be outdone.' James, John's father, did begin life as a servant in a gentleman's house in County Cork, where he met and married his fellow servant Anne in 1846. But within a few years of the wedding they moved to Cappoquin where James took a share in a draper's and haberdasher's shop.

John, born in 1850, one year after the end of Ireland's terrible famine, improved considerably on his father's achievements. He had little formal education but acquired manners and a good general knowledge through an aunt and uncle who were teachers. As a young man he became assistant to the engineer who brought the railway through Cappoquin, and later went to London to work for the South of Ireland Wheel and Wagon Company, which made the wheels for London's hansom cabs, and was sent by them to New York.

The Cappoquin he came back to had a population of some one thousand, little more than half of what it had been a hundred years earlier, and to earn a living he turned his hand to everything. The town was well-suited for trading, for it lies at a bend in the Blackwater where the river enters its tidal stretch as it turns south to Youghal and the sea. For a time he ran passenger steamers between Youghal and Cappoquin as well as schooners that brought coal from South Wales and carried away the produce of Waterford's farms and timber from the sawmill that he managed for Sir John Keane. His shop dealt in every sort of useful merchandise, and by 1909 Stanley and Co. was describing itself as general merchants, auctioneers, valuers, seed and manure merchant, draper and 'boot dealer'. He was a farmer, too, in a modest way, and was said to have 'walked his farm' every day.

In 1894, aged 44, he made a good marriage. Louisa Walker was the daughter of an employee of the Valuations Office in Dublin, a social notch above John's merchant father. The Walkers claimed to be descendants of the Rev. George Walker, the brave but intolerant Church of Ireland rector who, as governor of Londonderry, rallied the town's inhabitants during its siege by troops of the Catholic King James II in 1689, and died the following year, militant to the end, in the Battle of the Boyne. Louisa was also connected to another substantial Protestant family, the linen-making Orrs of Orrstown; Charlie, the Stanleys' second son who was sent ahead to prepare the summer camp site on the Cunnigar, was given Orr as his second Christian name. Louisa liked to say to her children 'there's as good blood in your veins as any in Ireland.' She herself had something of the sophistication of the gentry, and though described as 'very Irish in her ways' was a rather formal woman. Intelligent and firm, she ran an efficient household. There was usually both a maid and an assistant from the Stanley shop living with

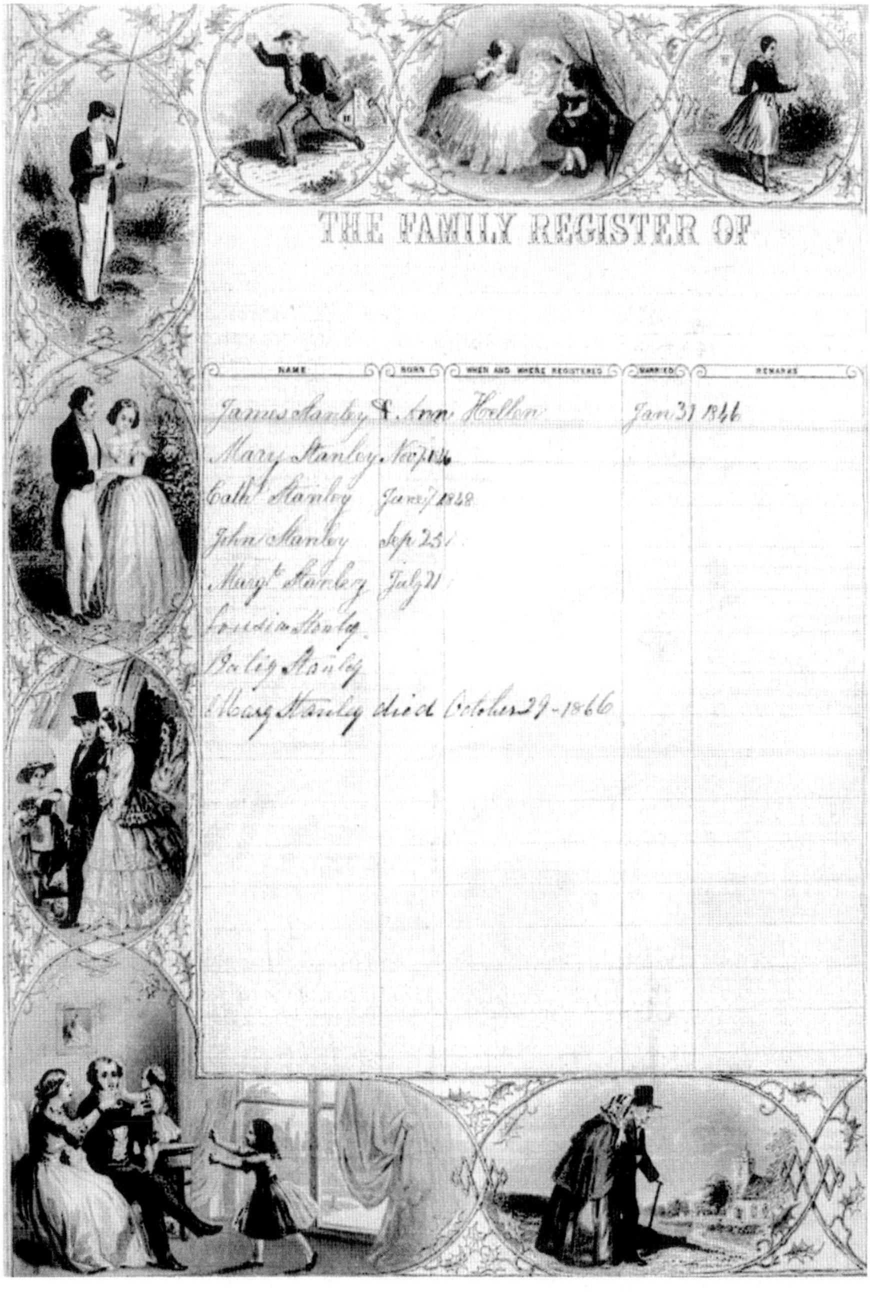

Figure 1.6   Stanley family register showing marriage of C.O. Stanley's grandparents in 1846, and the birth of his father, John

Figure 1.7   Letter to C.O. Stanley's grandfather in 1840 (prior to the introduction of postage stamps)

the family in the three-storey house on Main Street. Measured against the poverty of the Irish countryside at that time the Stanleys lived a decent, if modest, life.

Some members of the family felt that Louisa, the child of a government employee, did not quite approve of her husband's provincial and never very successful business activities, but if that was true it did not weaken the marriage. John Stanley had charm. As an old woman Louisa would sometimes say to a daughter, 'Jack had only to look at me . . .'

Her husband's greatest achievement lay outside his work. James Moore, the engineer with whom John worked to bring the railway to Cappoquin, was a champion sculler of Ireland. He taught his assistant to row, and rowing became the passion of Stanley's life. Under the guidance of Sir John Keane, founder of Trinity College, Dublin's boat club, Moore and John Stanley created at Cappoquin one of the most active

Figure 1.8   C.O. Stanley's parents, John and Louisa, c.1905

rowing clubs in Ireland. Stanley became a champion sculler himself and, in the words of the *Irish Times*, 'one of the great rowing figures of Ireland.' He rowed for Cappoquin until he was 40, when he turned cox and in 1911 coxed the town's Senior Four in their colours of dark blue to a famous victory over Trinity College itself. The triumph was recorded in a song written by a descendant of Daniel O'Connell, the champion of Irish Catholic rights. Sung to the tune of the Eton Boating

Song it included the lines

> We'll row, row, together
> To honour the 'Old Dark Blue.'
> Our veteran cox to guide us,
> We'll conquer the Varsity Crew.

Stanley became the town's chief rowing coach. He also taught all his children not only to row, but to sail, how to fall out of a boat, and how to save and be saved in the water. He put his four daughters in their own Ladies' Four, while Eddie, the eldest son, seemed likely to be as fine an oarsman as his father, at 17 rowing stroke in the Cappoquin Racing Four coxed by his 65-year-old father (later that year Eddie caught polio, which left him lame and ended his life as an athlete). John Stanley helped organise regattas all over Ireland, acquiring a knowledge of its waterways that was valuable to a 1907 Royal Commission on water transport. The Commission's chairman remarked that Mr Stanley seemed to know a great deal about Irish rivers. 'I ought to', he answered. 'I have rowed on every one of them and they are all beautiful.'

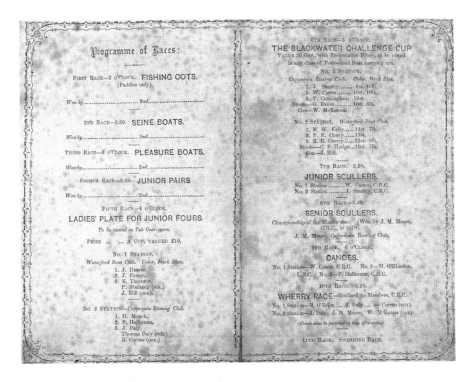

Figure 1.9   Programme from Blackwater Annual Regatta, 1880, showing C.O. Stanley's father, John, rowing for Cappoquin Rowing Club

*Figure 1.10   Cappoquin Senior Four, coxed by C.O.'s father, John, winning the
Subscriber's Cup at Dublin Regatta, 1911*

Raising money for his town's boat club brought Stanley into contact with the powerful local Anglo-Irish families: the Keanes, of course, but also the Devonshires, the Musgraves and the Villiers-Stuarts of Dromana, whose money built Cappoquin a fine new boat house just before the beginning of the Great War. But neither his skill on the water nor his position as president of the Cappoquin Boat Club could bridge the gap between these grandees and John Stanley, owner of the General Merchants' store on Main Street. Rowing had brought him into contact with this elevated world, but at the boat club Stanley coached ordinary Irish boys with names like Murphy, Heaphy and Curran, his neighbours' children. Rowing was a democratic sport in Cappoquin, and John Stanley did not take part in the gentry's own amusements, nor did he teach his children to. Eddie and Charlie did not go shooting. If any of the children learned to ride it never became their passion. None of them took up the great Anglo-Irish sport of hunting. Water was their element, and it welcomed any child in Cappoquin, however poor.

The Stanleys, Protestant and of modest means, were not well-placed in the Ireland of the turn of the century, where arguments for and against Irish independence ebbed and surged, usually as a result of British governments' mistakes and rarer successes in dealing with their Irish subjects. The great Protestant families could look after themselves: the Keanes would even become supporters of the future Irish

Figure 1.11   C.O. Stanley's father (centre, as cox) and the Cappoquin Senior Four rowers with the Subscriber's Cup, after their victory at the Dublin Regatta, 1911

Free State, with Sir John one of its first senators. But what could the Stanleys, who had no powerful connections, expect in a new Ireland? Cappoquin itself saw little of the violence that built up in the years leading to Irish independence. The worst the Stanley boys and girls seem to have experienced were taunts of 'Proddy-woddy green guts' from the Catholic children they lived among, and many of whom were their friends. John Stanley, engrossed in the Boat Club and rowing, might have been content with life in Cappoquin, but Louisa saw how limited her children's chances were, and she encouraged them to think of moving to England. Eddie remembered her telling them that 'if we didn't get out of Cappoquin, life would be very dull for us.'

Even without the difficulties brought by the struggle for independence the Stanley children's economic prospects were not good, but the family did not behave as though it felt itself under siege. The militant Protestantism of the Orange movement did not tempt them: how could it, given John Stanley's commitment to the community in which he lived? Both John and Louisa were good members of the Church of Ireland – their sense of identity depended on it – but their religion taught responsibility to the overwhelmingly Catholic world in which they lived.

John Stanley may have sought protection when he joined a masonic lodge in Fermoy, though the freemasons were themselves under pressure at this time. As long as the family remained in Ireland their best defence against an uncertain future was themselves. From an early age the children were taught to look after each other, and they constructed an intense, caring, squabbling world, its apartness reinforced by the knowledge of their difference from the other children among whom they were growing up. Their closeness found expression in their names. The parents gave the four girls conventional Christian names, but their aunts, one of whom ran the Cappoquin bakery, followed a fashion of the time and preferred to call them after flowers and precious stones. Louisa, the eldest, became Pearl. Her next sister, Ethel, was renamed Pansy, but no one liked that and she became Pan. Dorothea turned into Lily, which also did not take and changed to Ginger for the good reason that she had red hair. The youngest, Ursula, became Ruby, soon shortened to Rue. These were the names they used for the rest of their lives.

Eddie's name never changed, but stayed as constant as his good nature and good manners. As a child he fell in love with music, and was only 11 when the Stanleys' upright piano was carried from the house on Main Street to the Boat Club for his first public concert. He was clever, too, and won a scholarship to Trinity College, Dublin, a turning-point in the family's life. Charlie could not match that. Like his brother he was slight and short (neither grew to much more than 5 ft 8 in) but he was mischievous to the point of being wild. Pan, the sister closest to him in age, was often his companion in misbehaviour. They tied the door knockers of neighbours' houses together so that when a front door opened there was a rapping at doors along the street. They made a game of going uninvited to funerals, though that stopped after Charlie found himself alone in a room with a laid-out corpse.

Charlie did not flourish at the school in Cappoquin, where he put squibs in the tool kit of his form master's bicycle, and sometimes played

*Figure 1.12   C.O. Stanley (on left of picture) with his brother Eddie and his sisters,
Ethel (Pan), Louisa (Pearl) and Dorothea (Ginger), 1903*

truant. When Eddie started piano lessons with a teacher who lived some stations down the line from Cappoquin, Charlie insisted on having lessons too. They set out off together but when the train made its first stop Charlie would have got out if his brother had not stopped him. 'God, boy', he said to Eddie. 'You don't mean you really go to these lessons?' His parents sent him to an aunt who lived near a good school at Bray, just south of Dublin, but she was too kind-hearted to control him, and when he defied his father's wishes and came home for the Cappoquin summer regatta he was not made to go back. He took to education only when he was sent, as a last resort, to Bishop Foy's, a school in Waterford that offered subsidised places to the sons of Church of Ireland families. The headmaster was perceptive enough to recognise the boy's intelligence and know how to stimulate it. Charlie spent three years at Bishop Foy's and even won a Gold Medal for his studies, but it seems to have been his conversations with the headmaster that did most to awake an awareness of the affairs of Ireland and the challenges ahead of him.

The boy was also unlike Eddie in that his name did change. At some point people stopped calling him Charlie, and referred to him instead by his initials, C.O. His sisters said it was because he was the Commanding Officer, the one who always organised and gave orders. By some accounts Charlie was already C.O. as a boy, the acknowledged master of ceremonies during the holidays by the sea. It seems unlikely, though, that the conscientious Eddie would have given up the responsibilities of the eldest child, or that Pearl, a forceful character herself, would have caved in so early to her younger brother.

There is a story about Charlie, remarkable in its detail, that survived in the family perhaps because it seemed to foretell his future. One Sunday when he was at Bishop Foy's he went rowing by himself in Waterford harbour and saw a man with a woman and child in a motor boat approaching from the sea. Its engine stopped when it was still outside the harbour. The man tried to restart it, and when he failed beckoned to the boy and asked him to tow them into Waterford. All right, Charlie said, but it will cost you half a crown. This angered the man, who lectured him on the Christian's duty to help those in trouble, and then went back to his engine. Charlie continued rowing round the stationary motor boat. When the man again failed to restart the engine he called out to him that he could have his half-crown, but only if he towed them quickly to safety. Fine, said Charlie, but it is now growing dark and the new price is a pound.

The younger brother's independence and taste for adventure served the family well in the increasingly uncertain world around them. More than 200 000 Irishmen, Catholic as well as Protestant, volunteered to fight for Britain in the war against Germany that began in 1914, but all

Figure 1.13   C.O. Stanley as an RAF Cadet, 1917

hope that Irish nationalists might abandon their struggle for independence vanished with the Easter Rising of 1916 and the violent events it set in motion.

The war of independence that began in 1919 was the final blow to the Stanleys' chance for an ordinary life in Ireland. The following year Cappoquin voted solidly for pro-independence parties. Local Irish Volunteers who stored arms in a house not far from John Stanley's shop made several raids in the town, and on one occasion shot and killed two policemen. The younger sisters, returning to Cork from school in England, found themselves driving though a newly threatening Irish countryside. There is a story that even John Stanley, so well-respected in Cappoquin, was taken from the house by gunmen and threatened.

The three eldest children had already left Ireland by then. In the year of the Easter Rising, after graduating from Trinity, Eddie took a job at a school in the London suburb of Northwood, and in the usual manner of migrating Irish families set about finding somewhere he could bring his brother and sisters. Through a friend at his local church (he was as good a church-goer as his father and mother) he rented a house in Harrow, north London, for £40 a year. Pearl, who was also to be a teacher, joined him there. Charlie followed and for a short time taught maths at the same school as Eddie. As an Irishman he was not subject to conscription but suddenly, and without telling anyone in the family, he volunteered for the newly created Royal Air Force and became a radio operator. It was October 1918, just one month before the signing of the armistice that brought the Great War to its end.

Rue, the baby of the family, had a memory of Charlie that like many childhood stories have as much to do with the future as the past. It concerned those weeks of midsummer freedom on the sands of the Cunnigar when her brother, she said, would often rig the sail on their little boat and set off on his own. This was what Rue remembered, and wanted others to imagine: a small, solitary figure far out at sea in an open boat.

Sources

(For details of these and other chapter sources see Acknowledgments and Source materials sections.)
COS files: COS1/8/1, 1/8/2 and 1/9.
Michael Bell interview: Rue Tayler.
Books: Bell, *A Phenomenon of the Thermionic Age*; Sullivan and McCarthy, *Cappoquin, A Walk Through History*.

Chapter 2

Birth of a salesman

Eddie, Pearl and Charlie moved into No. 8 Sheepcote Road, Harrow, with one bed, two mattresses, a card table and a cheap armchair. Pearl thought the house was haunted until her brothers discovered a set of rusty servants' bells that sometimes worked free and rang for long-vanished parlour maids. When Charlie returned there from the Air Force in 1919 the solid Victorian house seemed set to become a nest of teachers. With Eddie and Pearl already at schools, what better way was there for their younger brother to establish himself in England? He had done well in his short spell as a teacher before joining the RAF. A reference from the school's headmaster praised his ability 'to take almost any subject, in fact far more than is usual', and called him as an 'excellent disciplinarian, . . . of strong character, and quite able to hold his own, and that pleasantly'.

Once the three elder children were earning a decent living, the younger sisters were able to join them in Harrow. As a final step their parents would come to London; in the meanwhile John and Louisa Stanley supported the family operation by putting a joint of Irish beef on the train at Cappoquin each Friday for collection the next day at Paddington Station.

Eddie helped his younger brother get into the Finsbury City & Guilds, a well-known technical college where he himself had taught for a time. The other students came from similarly modest families, and included many ex-servicemen who like Charlie took a speeded-up two-year course rather than the three years needed for a diploma. He studied civil engineering, and graduated with a mark of 62 per cent (the year's best was 76, the worst 55). He then joined Eddie at a school in Dalston but it is unlikely he meant to stay there. One day he told his brother he needed to slip away for a few hours – he did not explain why – and asked him to take one of his classes. Eddie agreed, but was horrified to find that his brother had not wiped the blackboard clean after a French lesson he had just given. What he had written was full of mistakes and as soon as Charlie reappeared the elder brother reminded him that the prudent teacher always cleans the blackboard when he has finished with it. Charlie was not interested. He had, he said, just been interviewed for a job, and got it.

He did not tell anyone what he was planning, nor could anyone remember him explain why he decided to go into advertising, a profession still in its formative stage. Engineering Publicity Services Ltd (EPS), the firm that offered him the job, worked mainly for the engineering world. Charlie's time at the City & Guilds seemed to suit him for that, but some of the firm's most interesting clients were in the new business of radio, and EPS's owner Frank Murphy himself later became a radio manufacturer. Radio had the glamour of a new technology, and it was radio that caught Charlie's imagination. 1922, the year he abandoned teaching, was a turning point for radio in Britain. The Marconi Company, the pioneer of radio, had been licensed by the Post Office to begin transmitting broadcasts in 1920, but complaints from other manufacturers of radio equipment led within two years to the creation of the British Broadcasting Company, financed by the industry as a whole, which took over the Marconi transmitter.

Here was a potential new mass medium to match and even outstrip the popular press that had been launched at the turn of the century. The Post Office expected to issue 200 000 receiving licences in the BBC's first year of transmission, and within just four years this grew to two million. America, which was leading the way, already had 15 000 retail radio dealers the year the BBC was launched, and two years later no less than 500 radio stations. A BBC programme of 1925 trying to convey the seductiveness of radio compared the thrill the first time a family switched on its new wireless to a 'wedding day and the first ride on a bicycle', an interesting pairing of sex and technology.

The new breed of manufacturers trying to provide the public with this thrill had received a boost during the war. The military's dependence on signalling, and later radio, brought increased demand for thermionic valves, the devices that made modern radio possible. The radio receivers that began to appear in the shops in the 1920s were able to tune in to a radio signal from a broadcast station, then detect and amplify it to convert it into audible sound. Thermionic valves were the key to this process. Vacuum tubes made of glass (and clearly visible inside pretransistor radio sets), valves worked by controlling the flow of electrons inside the vacuum from a heated filament (a cathode) to an anode (a positive electron). The more valves a radio had, and the better their quality, the greater its ability to turn distant signals into good quality sound.

Valve makers were therefore at the heart of the approaching radio revolution, and it was not surprising that EPS treated the Mullard Radio

Valve Company as a particularly valuable client. The company's owner Stanley Mullard was a famously rough diamond with scant formal education, but his success in the prewar lamp industry led the Navy to appoint him head of its valve-testing laboratory in Portsmouth with the rank of captain in the Voluntary Reserve. Postwar naval orders for the silica valves needed for radio transmission allowed Mullard to set up his own company, which also led the field in popularising the R valves used in early radio receivers. Stanley Mullard's factory was making a thousand of them every week even before the BBC went on the air.

When Charles Stanley started work at EPS as a 'mechanical accounts manager' it was already plain (as Mullard's success suggested) that radio's future lay with the valve even though crystal sets, the original radio receivers, were still being made by hundreds of manufacturers at a price several times lower than the cheapest valve receiver. The crystal set could not match the amplification of valves (which within a few years allowed the replacement of headphones by loudspeakers) nor their ability to select a frequency while avoiding interference from other signals. The 23-year-old Stanley felt so at home in this world that he asked Murphy to raise his salary. The latter refused, and when some time later he came across the young man with his feet on his desk and smoking a cigar a client had given him, fired him on the spot.

Charlie gave no sign that he was upset, and when he approached other advertising agencies demanded a salary of £1000 a year, unheard of for someone who had just started out in the business. Not surprisingly they all refused. He announced his next step one Sunday afternoon in the spring of 1923. He had gone for a walk round Harrow with Eddie and some of his sisters and they stopped to rest under the trees of an orchard. At first Charlie lay silently on the grass looking at the sky while the others talked. Then he told them, as though it was the most obvious thing in the world, that he was going to start his own advertising agency. At the age of 24, and with less than £20 in the bank, he proposed to become a principal in a profession in which he had only a few months' experience.

It was the end of Charlie Stanley, the most mischievous of the Stanley children, and the beginning of C.O., the man whose ability commanded the respect of those he grew up with quite as much as those who worked for him. C.O. set up the new business with a daring that would become his trademark, luring away from EPS an experienced account executive, George Royds, and Joseph Karsay, a talented Polish commercial artist. With Stanley as chairman they were the first directors of Arks Publicity, the name taken from the first letters of the three men's surnames with the A of Allard the office boy added at the front. The Art Deco logo on the office writing paper showed a rainbow curling over a floating ark.

Grand though it sounded, Arks was only a small room in Chancery Lane. Stanley took out £3 10s a week for his living expenses, and for a time he and Royds slept on beds in the one-room office, while Rue, C.O.'s youngest sister, came in each morning from Harrow to make their breakfast and was soon promoted, aged all of 16, to be the Arks secretary. The agency quickly picked up small accounts in the radio and technical fields but the breakthrough came when the great Captain Mullard telephoned to ask for a meeting. Rue took the call and when she told her brother he thought she was pulling his leg. The Mullard Valve Company, a major client for a well-established agency, would be a godsend for a novice like Arks. Royds and C.O. knew something of Mullard's demanding character from their days at EPS but at their first meeting managed to impress him enough to win the account. They celebrated by taking Rue for a steak and kidney pudding lunch at a Lyons Corner House, spoiled for her because she was afraid they would not have enough money to pay for the meal.

When less than a year old, Arks applied for membership of the Association of British Advertising Agents (today's Institute of Practitioners in Advertising). It listed six main clients of which the Mullard Valve Company and three others were connected with the radio industry. Arks gave its annual turnover as 'under £100 000'. It was probably less than half that amount, but it was already plain to one of the agency's young staff that 'we were in the thick of the most exciting new world you can imagine.'

A photograph of Stanley at this time shows a singularly composed young man. Sitting on the edge of a desk, he wears a three-piece light tweed suit and a silk tie, and holds a pipe in his right hand. His black hair is slicked backwards from a good forehead. The eyes (their blueness lost in the black and white photo) are large and serious, the mouth and chin firm. It is not an obviously handsome face, but it is compelling even if something of a pose. He was still young enough to want to buy, as an early proof of success, his first bowler hat at Lock, the smartest men's hatter in London, and to be, a friend from that time recalled, 'pretty uppity about this sign of sophistication'.

It was not long before what started as a pose became entirely natural. Certainly the handwriting of the man who signed the early Arks minutes suggested nothing but confidence. Those newly significant initials C.O. were written almost an inch high, and the cross on the t in the surname was even more pronounced. The extreme boldness of the signature was an accurate reflection of the balance of power, or rather lack of it, in Arks. When the agency was set up in 1923 Stanley, Karsay and Rhodes each held an equal share. Since the latter pair had considerably more experience than Stanley it might have been expected that Arks would be run as a triumvirate, but Karsay left within a year and at the beginning of 1925 C.O. made plain his distaste for shared authority. At a meeting of the Arks directors, the teenage Rue in attendance as acting company secretary,

> Mr Stanley said that unless the firm were prepared to allow him a full control over the policy and general direction of the firm he would have to resign his position. It was unanimously agreed that Mr Stanley should be given this control.

If the composed young man in the photograph might have been born, all of a piece, in postwar London, so Stanley's style of leadership seemed to emerge entirely with the first company he owned. Threatening resignation to extract recognition of his supremacy would become a favourite tactic. It worked not just because he was determined it should, but because those around him either recognised his greater ability or, as Royds was soon to do, left because they could not bear his domination.

The role of the Stanley family was particularly obvious in the fledgling Arks. Elder brother Eddie went to work there though he was quite happy teaching and, a concert pianist at heart, was never a natural businessman. Alec Guinness, just out of school and dreaming of the stage, worked at Arks for a while and was befriended by Eddie whom the would-be actor found 'kind, sort of amusing and very vague'. Eddie would play this role of a friend and comforter many times in a lifetime's work for his younger brother. Pearl joined, too; plainest of the sisters, but with the best brain for business.

With Rue this brought the total of Stanleys in Arks to four, and it rose to five when Ginger joined them. Not wanting his agency to be dismissed as a family outfit C.O. made Eddie disguise himself as Mr Walker (their mother's maiden name). Rue became Miss Walker while Pearl was turned into Miss Shaw, and would remain so for business purposes all her life. C.O. did not employ the family in order to spoil them. He never paid them generously, which was understandable when he launched Arks with so little money, but it became a habit. He would not let Rue take better paid jobs elsewhere, and when Pearl became pregnant with her first child he sacked her, deputing someone else to tell her the news (he took her back later).

The family put up with such treatment partly because working for C.O. was fun. There was the enjoyable conspiracy of brothers and sisters from somewhere unknown in County Waterford pulling off a bluff in London, the sense of childhood adventure prolonged as if by magic in the sophisticated world of British advertising. When the Mullard Valve Company was acquired by the big Dutch firm of Philips, C.O. also picked up work from them. In an attempt to increase sales of Philips' light bulbs in Britain he divided London into sections and sent Eddie, the sisters and their various boyfriends and husbands round shops to ask for Philips' bulbs, making them repeat their visits until

Figure 2.1   Eddie Stanley, C.O. Stanley's brother, c.1930

the storekeepers gave Philips an order. When he tried to improve his standing in the advertising community by taking part in the National Advertising Benevolent Society the family chipped in too. For several years he put on charity shows, usually light comedies, at the Scala Theatre in London. At Christmas 1927 Rue played the lead in *Paddy – the next best thing*. Another year Eddie and Pan starred in *Secrets*. Pearl, shrewd and sensible, looked after ticket sales, and C.O. got the kudos.

His own inventiveness and sense of mischief made him open to original ideas in others, and allowed him to appreciate two unusual talents who joined Arks when it moved to bigger offices in Lincoln's Inn Fields. Philip Zec came to the agency in 1929 as an immature teenage artist, and the Irishman William Connor as an office boy who, when Stanley saw what he was capable of, became a copywriter overnight. Both would become famous when they joined the *Daily Mirror*, Zec as a powerful cartoonist and Connor, using the pen-name Cassandra, as the unrivalled newspaper columnist of postwar Britain. Zec was 17

when he appeared before C.O. to be interviewed for the job of art director, and he quickly understood he was in the presence of someone who
'could charm the birds off trees'. Stanley praised the young man's work
and then asked if he had been to Mexico, New York or Paris, which
obviously he had not. Work for me, ran C.O.'s message, and you will
do the impossible; small wonder Zec could not resist him.

Figure 2.2   Front cover, designed by Zec, of Ark's portfolio of advertisements

Stanley did not mind if talented people were difficult and, by contrast, seemed uninterested in anyone and anything that was routine. Both Zec and Connor were left-wing, as were other young members of an office where, according to Alec Guinness, *Lady Chatterley's Lover* in a much underlined contraband edition and 'copies of *USSR Today* were passed from hand to hand, [the latter] greatly admired for its photographs of factory chimneys'. That did not worry C.O., who was never the slightest bit pink, any more than Connor's occasionally erratic behaviour and hard drinking. C.O. gave Connor his chance to write copy just because Zec, who had seen his work, told him the young man had talent. Zec thought that summed up C.O.'s 'genius': 'it did not matter who you were, whether you were 3 or 93, if he thought you could do what he wanted, he let you do it. He didn't interfere.' Connor's and Zec's method of work was to 'inspire' themselves with a couple of light ales, then sit down to write half a dozen rough slogans, sometimes starting from an idea thrown out by Stanley himself. They took these to him at the end of the day and 'the next thing we knew was that he had sold the campaign. Arks was the only agency I worked in where the clients would ring up and say we have to cancel the advertising, we can't keep up with the demand.' Zec did stylish, strong designs; among his favourites were a drawing of a London and North Eastern Railway train pounding through the night and, for Mullard, a picture of a fat man pouring tea into an overflowing cup because he is listening, entranced, to a new radio equipped with Mullard valves.

'When do you have time to sleep?' Zec asked him. C.O. said he slept all right, 'but I do sometimes go home and put my head in my hands and think'. Many years later Connor wrote 'what days they were in Arks ... I never learnt [sic] – and laughed so much in such a short time'. The respect between C.O. and talent was mutual, and was an early secret of his success.

The young boss had a less attractive side. His strength of character inspired awe as well as admiration, and that awe sometimes turned into fear. When charm did not work he was ready to get his way by shouting and thumping the table. An Arks employee who went on to a successful career elsewhere in advertising thought C.O. 'very rough on staff'. One way he 'terrified' people was by firing them if they came late to work, and later sending someone else, often Pearl, to tell them to forget it. He was so sure of his ability to make money for himself that he had little respect for the average man who had to put his family's well-being in the hands of an employer.

Talents such as Zec and Connor usually did not see C.O.'s harsh side, but Zec fell out with him in the end. Zec, who came from a large and poor family that needed his support, decided Stanley was a mean payer, though to do C.O. justice he warned his art director when he hired him, 'I shan't make you rich but I'll make you famous.' When

Figure 2.3   Front cover designed by Zec

Zec told him he was going to look for work that paid more, C.O. said he had a better idea. He had just bought an inactive company called Eagle Press (something he would make habit of for, as Zec observed, he was already making himself an expert in company law and finance). He offered to let Zec take all the Arks artists into the Eagle Press and at the same time do as much work for other agencies as he could manage.

C.O. would be his partner and they would split the profits fifty-fifty, Zec devoting himself to the creative work while C.O. saved his art director's valuable time by taking care of management. The young artist agreed, and when a year was up calculated that Eagle Press had made a profit of £8000, which meant a £4000 share for himself. C.O. said he would get the accounts drawn up. It took six months to complete them and they showed, according to Stanley, who was well aware of Zec's impoverished family, that their joint venture had made a profit of just £150. Zec noticed that the first item in C.O.'s balance sheet was £100 for 'depreciation of studio furniture' that was never more than second-hand bits and pieces. He lost his temper, demanded cash for his half of the meagre £150 profit – he wouldn't, he said, accept any cheque written by C.O. Stanley – and walked out of Arks there and then.

The meticulous, and to Donald Zec's mind, mean accounts drawn up for Eagle Press were scarcely typical of the way C.O. ran Arks, which was inspirational rather than ordered. Some of his employees made memorable mistakes. The young Alec Guinness marked a block for a Mullard advertisement in feet instead of inches, and was horrified when a taxi arrived with the monstrous result. As for C.O., he preferred his own instinct to market research, a science Arks ignored. The young creative staff might spend most of the day drinking coffee and then brain-storm a problem at the end of it. One of the copywriters remembered the office as 'undisciplined, unstructured chaos, but fun', the same word used by Rue to explain why she stayed with her brother in spite of the bad pay. Arks' style changed only when it merged in 1936 with W.S. Crawford and a more coherent management was imposed. When Zec left Arks and moved to J. Walter Thompson, a bigger agency, he got much more money but a great deal less responsibility and creative freedom than C.O. Stanley had allowed him as a matter of course.

C.O.'s employees cannot have realised how perilous their fun was. Started on a shoestring, and always undercapitalised, the agency's rapid expansion meant it was usually short of cash. Arks made profits, but they were erratic, no doubt partly because of general economic conditions. In 1930, with a staff of less than 40, the agency claimed a net profit of nearly £10 000; the following year, when the full effect of the 1929 Wall Street crash was felt, it managed only half as much. C.O.'s own salary as managing director was now £1250 a year, at that time a substantial sum. He was already perfecting the technique of bullying debtors and soft-soaping creditors, and his charm made a useful captive in Tom Eagle, his manager at Barclays bank. Eagle became a regular guest at the Arks annual dinner and took particular delight in C.O.'s Irish knack for speech-making.

Arks' unusual style led some in the advertising world to call it Sharks Duplicity. C.O. may have encouraged this when he set up a second agency to handle business that Arks, for reasons of good advertising practice, could not itself accept. Specialisation in the growing world of radio soon led to potential conflicts of interest, for an agency was not supposed to handle two clients that were in competition with each other. C.O.'s solution was E. Walter George, a new firm whose directors were Royds, the co-founder of Arks, and Eddie, still lightly disguised as Mr Walker. The impression given was that they had split away from Arks to run the new venture; the truth was that C.O. was its owner. This simple trick worked, and the two agencies were able to run in tandem such obviously competing clients as the battery makers Vidor and Ever Ready that Arks could not have kept on its books alone.

C.O. also won important clients outside the radio industry: the sock-maker Wolsey; Hercules bicycles; Phillips (not to be confused with the Dutch Philips), maker of soles and heels for shoes for which Pearl thought up the jingle 'Health and Leather go together'. C.O. also handled Rose's Lime Juice, though growing confidence in his marketing skills made him much more than Rose's advertising agent. The company asked for his help after it got into difficulties and was unable to pay a dividend. At that time Rose's concentrated on the sale of lime juice for industrial purposes, marketing it as a soft drink only in Jamaica, where it grew its limes. Stanley joined Rose's board and turned lime juice into a popular British drink with the help of a new advertising campaign and a new, and soon to be classic, bottle. He left the board five years later, when the company resumed dividend payments.

Radio brought the greatest opportunities. Most radios were still bought as parts, or as prepared kits, and assembled at home by enthusiasts who felt it was extravagant and faint-hearted to buy a ready-made set. The problem for a valve maker's advertising agency was how to persuade these radio enthusiasts to pick his client's product when there were so many competing components manufacturers.

Stanley worked so closely with the Mullard company that he later claimed credit for much of its success. He may have exaggerated, but not much, for he was appointed sales consultant to the Mullard Wireless Service Company that was set up to handle sales of Mullard valves, and for a time had an office at its headquarters where he thought up a daring scheme to popularise the Captain's valves. The many magazines aimed at home radio constructors offered competing designs, publicised the components of different manufacturers and were seldom written in a way to attract any but the most self-confident amateur. Stanley hit on the idea of putting out a magazine giving the design of simple sets that were easy to build, and specifying the components to be used mainly, but not exclusively, from Mullard. He planned a first issue with blueprints for four wireless sets, the cheapest costing only £2 15s, and all with the new and highly praised Mullard PM valve. The magazine would be sponsored by the makers of the components

Figure 2.4   Front cover of C.O. Stanley's innovative magazine introduced for Mullard

specified in the designs, and distributed through the radio dealers who sold the construction kits.

Stanley's preparation was thorough. The set designs were tested to make sure the recommended components were compatible, which in many other designs was not the case. Much of this work was done by a young engineer called Charles Harmer, who started off in the motor industry but, a passionate amateur set builder himself, switched to designing radios for construction magazines. Meticulous and cautious, he was the right man for the job.

C.O. was just as meticulous in the preparation of his marketing strategy. The name he chose for the magazine – *Radio for the Million* – and the language it was written in signalled that these construction kits were not just for knowledgeable enthusiasts, but within the understanding as well as the pocket of the average man more interested in listening to a radio than fiddling with its insides. C.O. christened the four sets described in *Radio for the Million's* first issue the Franklyn, the Rodney, the Nelson and the Grenville, names that were popular but classy and, since the last three were associated with famous admirals, even patriotic. The Nelson, with four valves the most expensive, was said to pick up 12 stations, continental as well as British, 'at good loudspeaker strength' while 20 more were 'good in the headphones'. The magazine made a point of reassuring those who were new to radio that they would be able to master the mysteries of set construction:

> We have set out to give our readers what must be recognised as a group of highly successful sets which, when constructed faithfully according to our simple instructions, and ultimately fitted with Mullard Master Radio Valves, will present Mr and Mrs Everyman with instruments capable of giving pleasing and endless musical entertainment.

The advantages of the Mullard valves were illustrated in the simplest nontechnical way by a picture of a man sitting on a swing suspended by the new valves' improved filaments: their strength was an important selling point because the filaments of earlier valves had caused set owners grief and expense by the speed with which they fractured. Stanley made sure the diagrams and blueprints showed the layout and connections of components clearly, and there was additional help from photographs of the different stages of the required construction. A £10 prize was offered to the reader who sent in the photo of the most neatly assembled set.

Radio dealers had an important part to play in Stanley's plan. He prepared a leaflet explaining the purpose of the campaign and how they could benefit from it. Chosen dealers who were to hold stocks both of the four recommended kits and copies of the magazine were sent publicity posters and suggestions for shop window designs. His most daring stroke was to launch *Radio for the Million* with a full page advertisement in the national press. The public largely ignored magazines written for the radio enthusiast, but they could not fail to notice an advertisement in the *Daily Mail*, the country's biggest circulation newspaper. The full page ad was less common then than now, and unheard of for a radio construction magazine. It showed the cover of *Radio for the Million* with its price of 1s and the slogan 'worth a shilling', but the trick was that *Daily Mail* readers who wrote to Mullard House would be sent a copy free, hence the advertisement's description of the new magazine as, 'This gift to the nation'.

The problem was to find the £900 that an advertisement of this size in such a paper cost. Although several components manufacturers put money into the magazine, Mullard was clearly its chief beneficiary, but the Captain was not the man to spend such a large sum on an experiment. Stanley persuaded him, only to walk into an even more difficult encounter with his client. The first issue of *Radio for the Million* appeared in December 1926. The *Mail* carried C.O.'s advertisement on a Saturday. Early the following Monday the Captain arrived at Stanley's office in Mullard House. He announced he had had a bad night, and was plainly irritable and apprehensive. Stanley left a description of what happened next:

> The first post arrived; it was small and contained not a single response to our great advertisement. Mullard was beside himself. He grew white and started to shout at me, standing up he cursed and swore at me, getting louder and shriller, and finally he wrenched the telephone from its moorings and threw it at me. At that moment, and just as it would have happened in a fairy tale, a Post-Office van drew up at the door. There had been so many replies to our advert that a special delivery had had to be made.

Radio for the Million won Mullard a quarter of the home radio construction market, and Stanley was soon more deeply involved in the project through Orr Radio and United Radio Manufacturers, two small companies he set up to manufacture construction kits for sale through the magazine. (United Radio Manufacturers was launched with a capital of £100, of which £90 was subscribed by Mullard, but these shares were later transferred to a nominee of Stanley's.) The campaign's success also won C.O. a secure place in Mullard's prickly heart. There were few people, Mullard said, 'who have so successfully interpreted [my] ideas and instructions' as C.O. Stanley, adding that Stanley was also ready to work at a lower than standard fee and without bothering to draw up a contract. He might have been defining C.O.'s way of doing business: first-class service, personal trust and sharp pricing.

Soon after the start-up of *Radio for the Million* Philips took complete ownership of Mullard's companies. Dutch directors were installed and two years later an unhappy Mullard resigned as managing director.

C.O.'s loss of such a supportive client was only partly compensated by his own relationship with Philips. Already involved in popularising new light bulbs for Philips Lamps, he had since 1926 been scouting out British companies for the Dutch firm to buy, and had even ticked off its managing director Anton Philips for rejecting a sales leaflet C.O. had designed for him. C.O. (or so it was remembered in Arks) said that while the Dutchman might know a great deal about valves and light bulbs he knew nothing about selling. The independent-minded Mullard was a character after C.O.'s heart, but the Dutch firm was a methodical bureaucracy, and he was soon in conflict with it. After taking control of Mullard Philips decided to hold a competition for the advertising account of Philips Lamps, even though Arks, on Captain Mullard's recommendation, was already its agent. C.O. only discovered this when he went to the Dutch company's London office to explain his plans for the next season's sales campaign. Instead he was asked to submit ideas for a trial campaign and told it would be judged against the work of competing agencies. Protesting that this was an insult to his professional competence he walked out. He saw his sister Rue not long afterwards and told her he had very likely lost the Philips business for good.

The indignant walk-out on grounds of injured honour was successful. Mullard, who was still managing director of his company at the time, came to Stanley's defence, complaining to Philips that their clumsiness had so upset this best of advertising agents that he was forced to prove his own trust in Stanley by giving him a formal contract, something neither of them had bothered with before. If the Dutch relented it was also because they understood C.O.'s potential value, and in 1929 NV Philips Eindhoven put him under contract to give 'advice and assistance for the increase of its turnover in the British Empire, especially as regards ... the market and possible ways ... for the increase of our selling activities'. Philips was to pay him £1500 a year on top of advertising fees, and an annual bonus of £500 if his results were 'at least ... equal to [those] obtained up till now', a reference to earlier consultancy he had done.

It did not change his critical opinion of Philips. In a letter to a Mullard director in the autumn of that year he complained that the Dutch had treated him 'in a most abused [sic] fashion', and accused a 'panicky' management of running Philips Lamps with 'enormous waste'. The same letter contained an assessment of his own worth and prospects that was as self-confident as it was shrewd. 'My real ability', he wrote, 'lies in sales cum advertising', repeating for emphasis that 'my great asset is sales analysis and constructive sales policy scientifically and economically conceived. It is only there that I am different to others in the advertising profession' (the claim to be 'scientific' sounded odd from a man who ridiculed market research; he seems to have meant there was method in his intuition). Arks, he went on, was 'an organisation unrivalled for what I want to do', and he intended to take on only 'that work which gives me ample protection for my staff and gives me the fullest scope and power to develop my personality and ability along the most congenial lines'.

The bullish tone falters in one place: 'I can deal with my own material problems – my domestic ones I shall never be able to.' Those few words suggest the anguish, perhaps self-reproach, of a man who has made a thorough mess of his private life, and understands at least some of the tricks his character plays on him. He had married in September 1924. The siblings were all getting married from Sheepcote Road and C.O., attractive to women and attracted by them, was never likely to be an exception. His wife, Elsie Gibb, was the daughter of a master tailor who also lived in Harrow. She was 29, four years older than her husband, and good-looking, with a better sense for clothes than C.O.'s sisters. A son, John, was born the following year, but the marriage collapsed not long afterwards. Elsie was good-natured, knew nothing of the world of business or what it might mean to be an ambitious young businessman's wife. She wanted a conventional domestic life, with a husband who came home each evening to put on the slippers she laid out for him. C.O. was too complex a character to coexist with simple good nature; cosy domesticity, if it ever attracted him, lost any appeal as his career took off. Elsie's sisters-in-law, who were nothing if not Irish, seem to have found her too primly English. There is no hint they much regretted the departure of their brother's wife.

Elsie, who seems to have adored her unusual husband, felt abandoned by the break-up. It disturbed C.O. too, for it was his first irreparable failure. In the world he was brought up in, the world of the Church of Ireland (to which he remained faithful) and his greatly respected parents, marriage was for life. For a while his work was not enough to occupy his troubled mind and he accepted a time-consuming job running the publicity committee of a Lambeth housing project organised by Lady Astor. He continued to support Elsie and John, who moved to a flat in Harrow. A young cousin who visited John there thought it a dreary household, redeemed only by the presence of an expensive German model railway, a present to his son from an increasingly successful, but absent, young father.

The future of both father and son was decided by a journey Stanley made to Cambridge in the spring of 1928. His interest there was the firm of W.G. Pye, a craftsman of Huguenot stock who joined the Cavendish Laboratory as an instrument maker in 1892, and set up his own business four years later. William Pye made a success of the venture, and World War I brought orders for military equipment such as gun sights and the first Aldis signalling lamps. After the war, partly

Figure 2.5   C.O. Stanley with his son, John, 1928

at the suggestion of Pye's son Harold, the firm entered the new world
of radio manufacturing, though without sacrificing the standards of the
precision instrument maker: when dissatisfied with the quality of components available on the market, Pye's radio team made their own. As
an undergraduate at St John's College, Harold Pye had been supervised
by a young don called Edward Appleton, later one of the commanding
figures of British science. Appleton was interested in the practical side

Figure 2.6   C.O. Stanley's first wife, Elsie, with their son, John, 1930

of radio and in 1924 helped his old pupil design a new and improved range of receivers whose success boosted Pye's radio business.

Stanley had been introduced to W.G. Pye by Mullard, who as a valve-maker cultivated potential customers among the setmakers. Arks took on publicity work for Pye, but Stanley was also interested in the firm in his role as talent scout for Philips. By 1927 sales of valve radios had overtaken crystal sets. It was not hard to foresee that, as production techniques improved and prices came down, manufactured valve radios would replace the kits and components that were bought by amateur

Figure 2.7   W.G. Pye (middle of front row) in 1900 with other members of staff at the Cavendish Laboratory

Figure 2.8   The W.G. Pye factory in 1923

home constructors. It was also probable that many manufacturers would either fail under the pressure of competition, or be gobbled up by bigger, stronger companies. In 1928 Britain had 107 radio manufacturers; six years later there would be only 43.

William Pye was still suspicious of the radio business, doubting people would want to listen to music chosen for them by others. If someone wished to listen to a Souza march, he argued, he would not want to hear anything else. Nevertheless he had turned down a proposal from Stanley that Philips buy both his firm's instrument and radio sides, and meanwhile Pye's radio business made steady progress under the management of T.A.W. Robinson, an enthusiast for the new technology in spite of his training as a Pye instrument maker. Thomas Robinson was a character in his own right. Workers joked that he had three brains because of the number of ideas he produced, and called him a 'tear-arse' for the way he rushed about the factory so that 'smoke [seemed to be] rising from the soles of his feet'. Robinson did see the advantages of the Stanley/Philips offer and was also well-disposed to a variant of it eventually agreed by Stanley and William Pye. C.O.'s new idea was that Philips would buy just the radio side for £60 000, and at the same time pay him a commission of £5000 for brokering the deal. Harold Pye thought it was too little, and complained that Stanley and Robinson had 'brainwashed' his father into accepting the offer during a lunch of the Radio Manufacturers' Association.

He was not the only one to question the proposed deal. This time Philips would have none of it, seeing no reason to pay Stanley commission for doing what they considered he was already under contract to do. That was why C.O. set off to Cambridge in the spring of 1928 with an offer of his own. He would buy Pye's radio business himself for £5000 down, the balance of the £60 000 to be paid after a delay of nine or ten months. In the meantime he would run the business, but if he was unable to come up with the £60 000 by the appointed date he would forfeit his deposit and everything would revert to W.G. Pye.

The instrument maker agreed, not knowing that Stanley did not even have the money to cover the cheque he wrote out then and there for £5000, though at least he had the prudence to hand it to Pye after the banks had closed. C.O.'s later recollection was that he had £800 in the bank, but it does not seem to have worried him. He did, though, telephone Tom Eagle, his manager at Barclays, to tell him what he had done. Susceptible though the banker was to C.O.'s charm the news alarmed him so much that he was waiting for him when he returned that evening to Liverpool Street station. Eagle delivered a lecture on the gravity of writing an uncovered cheque – Head Office would 'take a very dim view of it' – and said they were both summoned to appear the following morning before a representative of Barclays senior management.

C.O. guessed that G.F. Lewis, to whose office he and Eagle were conducted the next morning, was Welsh. He took heart from this, even though Lewis at once 'went up in smoke' and, in C.O.'s words, 'practically told me I was a criminal'. Lewis then turned his fire on the radio industry in general, declaring it was 'probably completely phoney' and likely to collapse overnight. Agitated, he got up and walked round his desk where he saw the bag that Stanley had brought into the office. 'What's that?', he asked. 'A wireless set', said C.O. 'It doesn't work', said Lewis. Stanley extracted the radio, put it on the desk and switched it on. There was silence, but when he rotated the set to catch the signal of the BBC's London station they heard music. 'Lewis had obviously never heard a wireless set before', C.O. thought, 'but as a Welshman at least he appreciated music. Ten minutes later, he promised he would do the best he could.' Other versions of C.O. Stanley's capture of Pye told by old friends and colleagues have him concluding his conquest of G.F. Lewis with a speech on how radio 'would change the lives of every man, woman and child in the country'. It was 'the most important invention they would ever see, the invention that would make its early pioneers vast fortunes'.

It was a salesman's victory. The radio he took to Barclays that morning in his new Lagonda was a Pye 25 set, one of the first portable radios in the country. 'Portable' in 1928 meant 20 inches high and 15 wide, as big as a travelling salesman's samples' case and, because of its large battery and accumulator, considerably heavier. It was, though, a handsome object, with a cabinet of polished mahogany, a neat handle on top, and a fretwork grill in the shape of a rising sun, Pye's Art Deco trademark. There was nothing odd about a senior manager of Barclays bank being amazed by a portable that unlike earlier radios had no visible aerial, batteries or loudspeaker. Shortly before C.O.'s fateful trip to Cambridge a young Pye engineer took a 25 set to London to test it for reception on the top of a double-decker bus, and watched with amusement as the passengers tried to work out where the music was coming from.

Stanley knew there was money in selling radios. That had been the chief purpose of Arks from the start. The question was whether he wanted to make them too. Workers at the Pye factory who spotted him on his visits to negotiate with William Pye thought it unlikely. Provincial businessmen like old Pye and Tom Robinson wore tweed jackets and grey flannel trousers. C.O. arrived in Cambridge in a smart suit, perhaps with his bowler hat from Lock. He did not impress the craftsmen, who under the finery saw a not very well-favoured young man, even 'rather a gorp'. To their mind, selling radios was one thing, making them quite another.

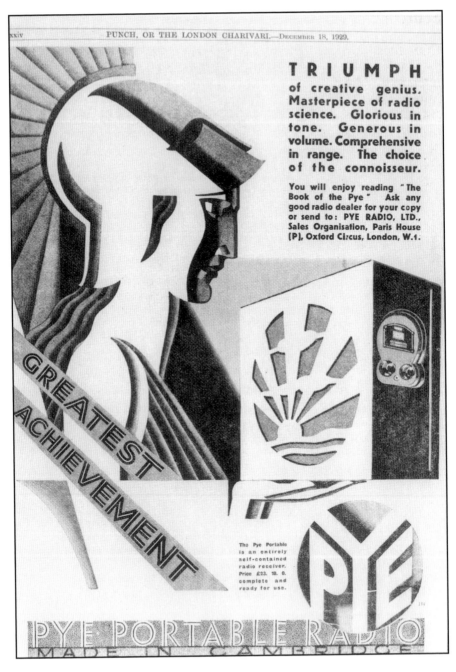

Figure 2.9

Sources

COS files: 3/1, 4/5 (Stanley's assessment of own abilities), 2/17 (Stanley's 1.9.66 letter to Michael Bell on purchase of Pye).
Interviews: Brian Beer, Fred Keys (purchase of Pye), Donald Zec.
Michael Bell interview: Rue Tayler.
A-Z files: Sir Alec Guinness. Fritz Philips (Stanley and Philips).
Books: Bell, *ibid*.; Geddes and Bussey, *The Setmakers*.

Chapter 3

Radio man

C.O. Stanley had every reason to suppose he had struck a good deal. The Granta works where W.G. Pye was making its radios occupied 27 000 square feet of land close to the river Cam in Chesterton, then a village on the Cambridge outskirts. New employees were struck by the clear layout of the plant and its cleanliness. Women, dressed in blue overalls and hats, were the majority of the workforce and did the assembly work; men operated the lathes and presses. As is often the case with things that are new, the factory hovered between the future and the past. When demand for radios fell off in the summer, women employees moved to the fields to pick fruit for Chivers, the East Anglian jam-maker. And while radio might be revolutionising the way people were informed and entertained, the cabinets of Pye radios remained a traditional craftsman's triumph of hand-polished wood (customers could choose between mahogany, walnut or oak). Some of the guarantee and service books in the light blue colour of the university were carefully inscribed by hand. Packing cases with completed sets were stamped 'Made in Cambridge – British and Best'.

If the cover of the factory magazine, the *Pyradian Gazette*, looked futuristic with a winged male figure throwing one arm round the globe and the other round the Pye logo, its contents suggested a simple, wholesome working life. There were reports on almost a dozen sports and social clubs, short stories, jokes ('this month's Scotch joke') and poems ('The jolly old Test Bench are we/ As happy, as happy can be'). The company gave a Christmas bonus of one week's wages, but did not pay workers when they were off sick; a weekly seven pence (7d) contribution to a 'sick club' assured an income during illness. Not everything at the Granta works was decorous. A young apprentice was shocked by the out-of-the-way store rooms where girls took the men they liked, and by the uninhibited behaviour of even senior staff at the annual works outing to the seaside.

Stanley's confidence in his acquisition was only partly explained by the quality of the factory and the radios it produced, for which much of the credit belonged to the tireless T.A.W. Robinson. He was also confident because he knew he could read a market and was sure, as he had explained to the doubting Mr Lewis at Barclays, that there was a rich demand for radios still to be satisfied. The BBC had now taken on its final form as the British Broadcasting Corporation – as opposed to Company – and was expanding transmissions. The quality of sets was improving; makers of what *Wireless World* called the 'junk' receivers that had made up a large proportion of radio output were going out of business. The industry's turnover would almost double between 1929 and 1931, for in spite of high unemployment following the crash of 1929 lower prices meant higher real wages for those who still had a job. Mains-powered sets, made possible by special valves, were shown by Pye and other manufacturers at the 1928 Radio Show. They were more convenient than sets depending on accumulators and heavy batteries that needed regular attention, and the popularity of the mains-powered radio was further boosted by the growing availability of electric power throughout the country (Britain's rate of growth of electricity consumption was the world's highest). BBC expansion meant that by 1934 most parts of the country could hear both national and regional programmes.

The technological breakthrough of the superhet (supersonic heterodyne) made it possible to design radio circuitry that, apart from the substitution of transistors for valves in the 1950s, remained little changed until recently. The superhet simplified tuning and made possible the greater selectivity needed to receive the growing number of radio stations, chiefly in continental Europe, whose broadcasts could now be picked up in Britain. New valves and circuits also solved the problem of local radio stations sounding shatteringly loud while more distant ones faded. Improved technology and standardised production helped the most efficient manufacturers. These were the firms that profited most from the surge in radio sales. The two million wireless licences issued in 1927, the first year of operation by the new style BBC, would rise to nine million by 1939.

There is little doubt that Stanley had a shrewd idea of the industry's golden future. Far less clear is what he meant to do with the radio company he proposed to buy. He certainly saw a chance to make money out of it, but perhaps as a maker of financial deals rather than of radios. He had already tried his hand at quick money-making, buying property in London that he at once resold at a profit. In the winter of 1928 he made his first visit to a still booming New York and on his return in the New Year launched Pye Radio as a public company. The offer, issued through the Gresham Trust, was constructed to ensure him control of the new company by dividing its capital between £150 000 Preferred Ordinary Shares at £1 each and 600 000 Deferred Ordinary Shares costing one shilling. This put a value on the company of £180 000, almost three times the sum Stanley had undertaken to pay W.G. Pye, proof enough of the excellence of his bargain. Preferred Ordinary Shares were popular in the 1920s because they were thought to offer a good return on capital, in this case a guaranteed dividend of 8 per cent rising to a maximum 10 per cent if profits allowed. Holders of Pye Deferred Ordinary Shares would be entitled to all profits above £15 000 and because of their greater number would control Pye Radio.

The prospectus laid down that Stanley's consideration for selling the company that he so tenuously owned was £82 100 in cash and £18 500 in one shilling Deferred Shares, i.e. 370 000 shares. He also subscribed to 30 000 £1 Preferred Shares which also entailed buying the same number of Deferred Shares at a cost of £1500. The result was that, while the shares he owned were worth £50 000, or less than a third of Pye Radio's capital, his 400 000 Deferred Shares, each with a vote, gave him the power of the majority shareholder. By the time he had paid William Pye the balance of the agreed purchase price and completed his own purchase of Pye shares, he was £16 000 out of pocket, scarcely a problem for a man with collateral in the shape of his new Pye holdings and who was learning to tease money out of banks as deftly as an angler coaxes fish from a river.

Pye Radio did not disappoint him. In its first three years it achieved an average annual profit of £82 000, from which Stanley received average annual dividend payments of almost £22 000, not bad for a man who in 1928 had only a few hundred pounds in the bank. He attended Pye's inaugural board meeting but then stayed away, having presumably chosen not to be one of its four directors. The most important of these were the joint managing directors, Robinson and R. Milward Ellis, a radio man well enough known in the industry to be elected chairman of the Radio Manufacturers' Association in 1931. Robinson was to look after production, while Ellis took care of sales. The attractions of the Pye chairman Sir Thomas Polson were not so obvious. In his sixties, a member of the Cavalry Club and author of a book on *Horses and Horsemanship*, Polson's appeal to Stanley may have been that he, too, was born and brought up in Ireland. He had also served as Chief Inspector of Clothing at the Royal Army Clothing Department and been, briefly, a Member of Parliament, and as a former chairman of Rolls Razor was an unexceptional figurehead for a new public company. Stanley's closest ties were to the fourth director, L.G. Hawkins, a competent businessman whose electrical appliance distribution company was a client of Arks.

Pye's chief shareholder still showed no sign of wanting to get directly involved in the business of making radios. The letter to the Mullard director in which he revealed his conviction that he was above all a salesman and master of the market was written several months after the launching of Pye Radio; he signed his five-year contract as a well-paid consultant to Philips just two months after Pye's launch. And the early 1930s were busy years at Arks. True, he had set up the two small radio companies of his own, United Radio Manufacturers and Orr Radio, but this was at Mullard's request in order to meet the demand for radio kits created by *Radio for the Million*: Mullard knew the market for kits would soon come to an end and found it convenient to have Stanley take care of it in its dying days.

Nor was it obvious that Stanley could have been of much help to Robinson and Ellis, both of whom knew what they were doing. Late in 1931 the pair went to the United States where they met industry leaders at RCA, Atwater Kent and Philco, and came back with information on the value of conveyor belts in chassis production and also circuit diagrams of all the current American models. By 1933 the Cambridge factory had grown to five times its size of 1929, and was turning out 40 000 receivers a year. Pye's new models became a feature of each Radio Show: the Twintriple Portable Receiver in 1930, the Q set in 1931, and the MM Transportable Receiver, which quickly won a reputation for reliability and good performance. Successive annual general meetings heard Polson praise the joint managing directors as men of 'very unusual ability', and the company's ranking as one of Britain's best medium-size radio companies seemed assured. (EMI with its Marconi and HMV trademarks dominated the radio market throughout the 1930s, followed by Ekco. Pye belonged in the next group with manufactures such as Murphy, Ultra and Ferranti.)

To those who worked in the radio industry Stanley seemed far removed from Pye's affairs. No one watched Pye more closely than Mullard, which, as Britain's biggest manufacturer of valves, was keen to make customers of Robinson and Ellis. But in spite of Stanley's close relationship with both Mullard and its Dutch parent company, the valve-makers were not sure what influence he had over the factory in Cambridge. In June 1933 an obviously frustrated S.S. Eriks, Mullard's general manager, informed the Philips head office in Eindhoven that he had told Stanley 'that if his advice is of benefit to [Mullard/Philips] he should at least be able to show it with the people [at Cambridge] with whom he is particularly friendly' – in other words he should be able to get Pye to buy Mullard valves. Eriks doubted that this friendliness amounted to much, suspecting that Ellis and Robinson 'do not always take Mr Stanley as seriously as he thinks they do'. Pye's joint managing directors, Eriks discovered, did not keep Stanley fully informed of their production plans and 'as a rule he does not actually see the new models [from the Cambridge factory] until they are ready for sale'. Stanley had himself admitted to Philips that he could not persuade Robinson to buy valves from Mullard or Philips because of a disagreement with

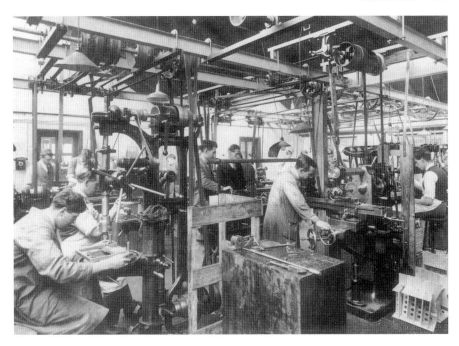

Figure 3.1   At Pye Radio Ltd., Cambridge, 1929: Tool room

Figure 3.2   At Pye Radio Ltd., Cambridge, 1929: Production and drilling shop

Figure 3.3   At Pye Radio Ltd., Cambridge, 1929: Coil-winding shop

Figure 3.4   At Pye Radio Ltd., Cambridge, 1929: Component assembly shop

Figure 3.5   At Pye Radio Ltd., Cambridge, 1929: Staining and filling shop

Figure 3.6   At Pye Radio Ltd., Cambridge, 1929: French polishing shop

Figure 3.7

Figure 3.8   At Pye Radio Ltd., Cambridge, 1932: The factory

Figure 3.9   At Pye Radio Ltd., Cambridge, 1932: Drilling, tapping and milling machines. Note the caps. These are worn to protect the hair from the machinery

Figure 3.10   At Pye Radio Ltd., Cambridge, 1932: Press section

Figure 3.11   At Pye Radio Ltd., Cambridge, 1932: Copper, cadmium and nickel plating shop

Figure 3.12   At Pye Radio Ltd., Cambridge, 1932: Coil winding

Figure 3.13   At Pye Radio Ltd., Cambridge, 1932: Component and part assembly shop

Figure 3.14   At Pye Radio Ltd., Cambridge, 1932: Assembly line with feeder benches

Figure 3.15   At Pye Radio Ltd., Cambridge, 1932: General assembly shop

Figure 3.16   At Pye Radio Ltd., Cambridge, 1932: Final inspection and cartoning shop

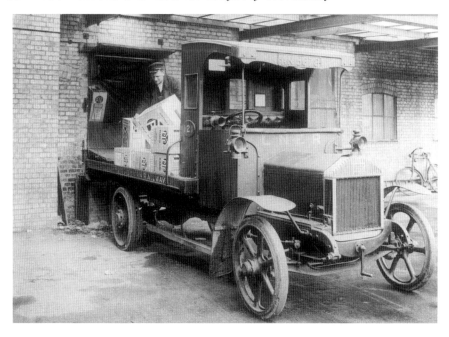

Figure 3.17   At Pye Radio Ltd., Cambridge, 1932: Down the chute and on to the LNER lorry

Figure 3.18   At Pye Radio Ltd., Cambridge, 1932: The end of a day's work

Figure 3.19

the Dutch firm going back to the days when Robinson was running the factory for W.G. Pye.

Eriks nevertheless appreciated the advertising Arks created for him ('there is no advertising agency in London which can carry out our work so well as [Stanley's] specialised staff'), even while wondering if Philips had got too 'friendly' with this puzzling figure, and allowed him to acquire too much expertise at their expense. On balance he thought it prudent to keep on good terms with a man who seemed to be acquiring 'some influence' in the Radio Manufacturers' Association by running its benevolent fund with the same flair he had shown earlier in his charity work for the advertising profession. And Eriks reassured himself with the thought that there was less than met the eye in a man who, though 'admittedly a somewhat extraordinary individual . . . usually defeats his own object as he is not sufficiently consistent'. Eriks thought the rest of the radio industry shared his doubts, noting that in spite of his charitable activities Stanley 'has no friends amongst the manufacturers who do not like or trust him'.

No one seemed to know what was he up to. A radio journalist described him as an 'unknown quantity, associated with everything and tied down to nothing'. How 'tied down' was he to Pye? Not very firmly, in the eyes of David Sarnoff, head of the Radio Corporation of America, who in 1932 offered him the job of managing director of the British arm of HMV. Stanley refused, citing his links with Philips, but he did not mention Pye. Later he told friends he never meant to run Pye himself; and that he still planned to sell it on to Philips after launching it as a public company. That, he said, was why he kept away from Cambridge and did not take the seat on Pye's board that could have been his. S.S. Eriks certainly thought he still wanted to link Pye with Eindhoven, believing 'his interests are best served by linking Pye with our group without necessarily wishing [us] to buy at the moment'.

Two events brought Stanley's scatter-gun life into focus. He persuaded Pye to revolutionise the way it sold radios, only to find the immediate result was an unnerving collapse in sales. And he fell in love.

On 1 July 1933 T.A.W. Robinson announced that Milward Ellis had suffered 'a breakdown in health and had gone abroad', and that C.O. Stanley had agreed to join Pye as a director (the list of 14 other companies C.O. was obliged to declare an interest in suggested the range of his interests: apart from Arks and other publicity-related ventures there were six radio/electrical companies, two life insurance firms, and the Irish textile makers Sunbeam Wolsey). Pye's results the following year explained the urgency behind his move. Profits for the year ending 31 March 1933 had been £108 000 (adjusted for inflation, the equivalent of £3.7 million in the money of 2002). The following year profits sank to £51 000, and it was 4 years before they reached £100 000 again. The crisis had become apparent within weeks of Sir Thomas Polson praising both Ellis and Robinson to the skies at the 1933 annual general meeting.

Pye's records make it look as though C.O. Stanley, the master salesman, agreed to join the board to help the company in its hour of need.

The truth was that Stanley began to reorganise the Pye sales system several weeks before he became a director of the company. Pye had always dealt with a small number of wholesalers who sold the radios on to retail shops, but by 1933 other radio manufacturers were switching to selling through retailers. This gave them better control over how their sets were sold and was also more profitable, for wholesalers could demand bigger discounts than any retail shop; Stanley also saw selling through retailers as the way of the future, and switched Pye's distribution to 3000 retail dealers on the same day that T.A.W. Robinson told the board of Ellis's 'breakdown' and the invitation to Stanley to join the board.

Polson later admitted that Pye knew little about the newly appointed dealers on whom its fortunes now depended. As for the jilted wholesalers, they dumped their stocks of Pye sets as quickly as they could, thus spoiling the market for the company's new retailers. In the first two months following the switch, sales were down 30 per cent on the year before. The situation only began to improve when Pye brought out new models that were exclusive to its new retail network.

Polson's remarks suggest that the operation was carried out with more haste than care. Stanley himself admitted that 'within twenty eight days [of the switch] every channel of distribution was closed to the company', and that it was five years before he managed to indoctrinate radio shop owners so that they 'saw Pye dealerships in their dreams'. Certainly the watchful Eriks questioned the way Pye performed its tricky manoeuvre. He knew Stanley had also urged Philips 'to go to the dealers direct', and thought his purpose had been to nudge the Dutch 'to pave the way for Pye to do the same'.

Eriks had no doubt that Ellis, whom he called the 'father' of the system linking radio manufacturers and wholesalers, thought Pye's change in trading policy dangerous, not least because of the way it was carried out. Eriks himself believed it a hasty decision rather than 'a properly conceived scheme with provisions for continued distribution in certain outlying parts of the country'. Eriks' conclusion was that by forcing the new scheme on Pye, Stanley had damaged Ellis's prestige in the industry. The upshot was that Ellis, whom Eriks considered 'a very independent person' – and certainly not ill, as Pye claimed – could not stand the 'humiliation', and chose to let Stanley put the dubious plan into effect. When sales collapsed, Stanley had no choice but to involve himself more deeply in Pye to protect his name as much as his financial interest, for there was already speculation that if Pye's new scheme failed Milward Ellis would conveniently turn up to restore the old way of doing things.

S.S. Eriks thought Stanley had taken the risk in the first place because 'he wanted something to concentrate on, and so conceived the dealers' scheme in which he intended to take an active part as a promoter'. This wish to absorb himself in one big project had much to do with Velma Price, the Australian Stanley first met in London in 1932 and with whom he almost at once fell in love. Some of his friends were surprised to find Velma so 'young and pretty', but she was also intelligent, being one of the first women to graduate from Melbourne University's medical school. The relationship developed during the sales crisis at Pye, and from Stanley's letters to her it is plain he had found what his first wife failed to be: a self-confident woman who could take an interest in his world of business, and an equal – Velma would not settle for less – he could trust.

It was a passionate affair. When Velma went back to Australia for some months towards the end of 1933 he comforted himself by imagining how on her return he would move 'your hat on your hair over your right ear and kiss it so hard, then you'd know it was really me'. In another letter he imagined her as a 'Polynesian with hair that I would love to run my fingers through and powder I used to put on your back and . . .'. The writing was often incoherent, for he felt wretched without her. 'For no reason, for all the reasons possible I have surrendered myself wholly and completely to you . . . if I loved you less it might be better.' The memory of an evening when he had first seen her in a particular white frock 'still lives and shorts right inside me and has made me at times the most miserable creature'.

He worried that he might have hurt or changed her. 'Darling, say I haven't. Say you'll just be you again. I want to see you laugh – go mad – cry (real hard) – be you.' In one of her letters she called him 'a sham' and, though he protested he could not live without her, he admitted he was 'afraid to say the things I want to say most in case I was wrong and didn't mean to implement them'. Velma played with him, rebuking him when his letters lacked passion, and testing him with hints that she might not come back to England. Unaccustomed emotions put him in a mood for analysis, of himself as well as Velma. When he wrote that she was 'the one woman in the world that means the smallest serious item in my most selfish life' it was perhaps the lover's traditional confession of unworthiness, but he followed with a promise of reform. 'I'll try to be less selfish . . . the trying will be quite natural without having to make it so.' But he also apologised for his nature which he described as 'twisted and queer' as well as selfish, and it was plain he saw in her a character just as unusual. He reminded her that she had said that 'convention [did not] exist in anything we want to get out of life', just as he himself did not care 'two knockers in hell for all the public opinion conventions'. Their problem was to work out how two such potent personalities might coexist in marriage:

> We both realise that we have a difficult job to keep our temperaments welded together. Although both of us are in the big issues of life far

removed from the selfish, along a small number of other paths we are terribly selfish-flavoured with a good sprinkling of the egoist. Always running the risk we may not realise where we are wrong – in time to put it right.

He admitted he had been 'told off so long for ruling the lives of others that with anyone I love – it's even more difficult'. They had both shown their inclination to dominate others by making plain their disapproval of Rue's marriage. It was, for him, an old story. 'The problem is if one thinks for other people, it is always by your *own* foot rule.' This was why he had also opposed the marriages of Pearl and Pan, and only now realised he had been wrong. He seemed to feel Velma would encourage rather than restrain his inclination to dominate, but that at the same time he would have to fight his ground against her. In one letter he worried about men who married 'strong and attractive women' only to find friends pigeon-holing their wives as 'damn self-willed' while condescending to the husbands as 'dears'. Was this why he slipped in a reminder that she should not think he was always wrong?

He wanted a woman he could talk to, and since by chance and choice he found his satisfaction as a businessman she had to have the taste and mind to understand that world too. There were times, he told her, when all he wanted was to get away from his work and be alone with her. But he also knew that without his work he would be nothing: 'if I did not have something to attain and talk about I would soon hate life itself'.

Within a year of meeting Velma Price, Stanley was explaining to her his moves at Pye in the months after Milward Ellis's unconvincing illness. When Ellis went abroad, supposedly to convalesce, Stanley moved into his office at Pye's London headquarters in Africa House on the Strand and took control of the company's selling operation. The state of relations between the two men may be judged by Ellis's behaviour when he returned to Britain. He waited for a week before agreeing to meet Stanley, and then failed to keep the appointment. The trouble with Ellis, C.O. told Velma, was that he lacked nerve, a hard judgement but the saving of Ellis, for as long as C.O. felt he was not to be taken seriously as a rival he could afford to let him stay at Pye. He could not bring himself to like Ellis, though, and when the latter moved into a new house in Cambridge and gave a house-warming party, C.O. told Velma, who was in London, that she was lucky to have an excuse not to go.

He also took care to make sure Ellis should not attempt a come-back, in the early autumn of 1933 making a thoroughly disingenuous declaration to the Pye board that he proudly recounted to Velma: 'There is a small thing that worries me', he had told his fellow directors. 'Ellis and I are great friends. I have sort of arrived in this business without the required formality, and it might be easier to handle a difficult position if we talked about it.' He got what he wanted, explaining to Velma that 'I have taken over the commercial side and [Ellis] will busy himself with politics, patents etc. and not even come into the office except to keep particular appointments'. *Busy himself with politics, patents etc. –* the contempt could not be plainer.

C.O.'s colleagues apparently accepted his claim that he had 'taken the firm through a real crisis to leave it sounder and better than before'. At the AGM the following year Polson would declare that 'credit for this restored happy state of affairs' went mainly to C.O. Stanley. Pye owed him 'more than words can express', and was lucky to have the services of this man 'with an international reputation in marketing'.

In fact the battle for the control of Pye went on, and a pact reached at the end of October 1933 under which Ellis, Robinson and Stanley would run the company as a committee of joint managing directors only served to trigger C.O.'s next move. 'We decided to run things as a committee, Ellis in the chair', he wrote scornfully to Velma, but 'at the board meeting to put it over I threw in my hand and said I did not believe in any job being run by a committee'. He described how he had browbeaten the board: 'got terribly worked [up], nerves to bits, banged the table. Old Polson nearly died of heart failure'. After this performance he took Ellis and Robinson to a hotel where 'they agreed to run [Pye] without any committee [and] leaving the whole commercial side to me'. A few days later Ellis collapsed at his golf club, though he recovered later.

In spite of C.O.'s objections management by committee lived on for several months for in November they again agreed that Pye should be run by a 'management committee' of Ellis, Robinson and Stanley. C.O. gave his approval, but on the day they were supposed to put their signatures to the deal he again went back on his word. 'On the strike of the clock I refused to sign the proposed agreements so we are back where we started. But my sweetheart don't worry – I'm losing all my inhibitions.' He obviously felt he had Velma's approval for his increasingly brutal tactics.

The new problem was not Ellis, but T.A.W. Robinson, with whom Stanley had made friends in a way he never bothered with Ellis. Robin, as C.O. called him, had taken his side in the tussle over the switch to selling through Pye-appointed dealers, and together with Polson helped Stanley prevent Ellis take back control of Pye's commercial operations. Now that C.O. planned to marry Velma, Robinson was helping him with the preparations for his divorce from Elsie. C.O. had agreed to let her sue him and, to meet the demands of the divorce laws and to keep Velma's name out of the courts, had to provide evidence of a supposed liaison with another woman. Robinson brought round to C.O.'s flat in Wilton Crescent Mews the 'roughneck' – C.O.'s description – private detective whose job was to provide this evidence.

C.O. confessed to Velma that this was harder to arrange than he expected, and 'if in my endeavours to meet my own taste (notice that first) my scruples and your very pointed wishes I don't become a woman-hater I'm not the man I thought I was'. The 'hateful' business was accomplished, and the divorce became absolute on 26 November 1934. C.O. and Velma married three days later at Westminster Registry Office with Robinson and his wife as witnesses. In the marriage certificate Stanley described his father as a 'gentleman farmer'; Velma called hers a 'grazier', which was true as far as it went, though he also trained race-horses, a dubious occupation given Australian racing's then links to the criminal underworld. The married couple set up house in Lowndes Place, Knightsbridge. C.O.'s old life, in which he moved constantly around the country meeting clients and making speeches – as the journalist put it, 'associated with everything and tied down to nothing' – was over.

As much as a year before the wedding the well-informed S.S. Eriks was predicting that Stanley would find Robinson 'rather an obstacle' and most likely 'clear him out of the way'. Robinson may have seemed a formidable character to employees at the Granta Works but there is no evidence he was ever a match for C.O. Stanley. His strength was as a maker of wireless sets, but by the autumn of 1933 Stanley had managed, in his own words, to 'put in' a man at Cambridge who discovered 'a great deal was wrong on the business side'.

The evidence of C.O. Stanley's ten-year career in business was that he could not tolerate equals. Within months of setting up Arks he demanded and got the consent of his two more experienced partners that he should take charge of the business. Robinson might know more about radios, but it was C.O. Stanley who was becoming a celebrity in the radio world. Within days of his appointment to the Pye board in 1933 the magazine *Wireless Trader* declared that in an industry full of romance 'there was no greater romance than that of C.O. Stanley'. Early the following year *Radio Magazine* expressed surprise at his 'unexpected oratorical powers'; the radio industry 'has been languishing for a lack of leadership; it need languish no longer'.

Robinson's position at Pye was further undermined by the direction in which the industry was moving. The battle in the 1930s was to sell this new instrument of entertainment and education to ever wider, and therefore less well-off, sections of the public – two million of the nine million radio licences of 1939 were bought by people with incomes of less than £2 10s a week. Though buying sets on hire purchase was becoming more common, radio ownership in Britain still lagged behind the USA, where families no longer used speaker extensions when they wanted a radio in another room, but bought one of the increasingly inexpensive midget sets. The Ullswater Committee, appointed in 1936 to decide the future of the BBC charter, criticised the industry for pricing its radios too high, and suggested that manufacturers collaborate with the BBC to produce a standardised and more affordable set. This did not happen, but that year both Philips and Philco (a subsidiary of the American Philco Company) brought out radios that could be bought for only 6 guineas. Philco called it the 'People's Set', but Philips achieved the lower production cost of just £2 10s for each receiver. With prices under such pressure (Polson complained of 'cut-price competition') it was perhaps surprising there were still 600 different radio models on the British market in 1937.

Ask connoisseurs of vintage radios what they think of the radios Pye made in the latter part of the 1930s and the answer is likely to be that they lack the quality of the sets produced when T.A.W. Robinson ran the factory. In these later sets wood veneers replaced solid hardwood in the cabinets and Pye's well-known fretwork motif of the rising sun was scrapped, though Stanley would not follow Ekco's example and make truly modern cabinets out of moulded bakelite. Pye also brought down costs by reducing the number of components in each set. The historian of radio Keith Geddes told a story of Stanley walking through Pye's research laboratory at Cambridge and stopping at the bench of an engineer working on the chassis of a new set. He pointed to a resistor and asked what it was. The tone correction circuit, said the engineer. Stanley told him to cut it out. The radio was switched on and there was no apparent difference in sound. Accusing the engineer of putting in unnecessary components C.O. pointed to the capacitor. It too was removed, again with no effect on sound quality, which was not surprising since the resistor was already gone. You're wasting my money, C.O. scolded the engineer, and ordered him to remove the main electrolytic condenser. Impossible, said the engineer. Take it out, said Stanley. The condenser went, and only when the radio produced a loud hum did C.O. admit that it had to go back. Never, he said to the engineer before leaving, never use unnecessary components again.

Cut-price methods help explain why Pye, which produced only 30 000 radios each year at the start of the 1930s, was turning out 200 000 a year by the end of the decade, and achieved record profits of £123 000 in 1937. By 1936 Pye was able to put on the market a basic battery-powered portable radio that, at £7 17s 6d, cost little more than the Philco 'People's Set'. Retailers' publicity for the Pye T70 admitted it could not pick up as many British and continental programmes as the company's more expensive sets, but called it ideal for people who wanted a second set in the house or only tuned in occasionally.

Stanley had quickly understood that, to be a commercial success, radio had to be sold to people who neither knew what was inside their set nor demanded the highest possible performance from it. Compared to the cheaper Pye sets the radios produced by Murphy (run by

C.O.'s former advertising boss Frank Murphy) were more handsome as objects, and technically superior, but Pye was flourishing at the end of the 1930s while Murphy was not. Stanley also understood that if radios were to become truly popular they had to be popularly advertised. Set-makers traditionally advertised in the trade and technical press, but Stanley began to place publicity in widely read magazines such as *John Bull, Radio Times* and *Picture Post*. When other manufacturers followed suit they usually did not do it so well.

In 1936 Pye launched the QAC 2. At 8 guineas it was the cheapest AC set then in the Pye range and was planned, Stanley told the company's sales representatives, as the first in a series of new radios that would appear each year and give the best performance for the lowest price. This was his recipe for overtaking such market leaders as Philips and the EMI trademarks Marconi and HMV. The QAC 2 became Pye's best-selling radio, and its sales kept the Cambridge factory fully loaded throughout the traditionally slow summer months. At the same time C.O. brought in a new rental scheme to allow Pye to fight back against growing competition from rental companies, though he always maintained that a good salesman could always persuade a customer to buy rather than rent. He also launched a £12 000 four-week advertising campaign in the national press under the slogan 'Switch to Pye', which he promised would set the whole industry talking.

Stanley kept up relentless pressure on salesmen and dealers to push Pye sets, particularly to the many people who owned poor-quality radios that he claimed gave only 'half the [potential] entertainment value' of a decent set. Even the troubled summer of 1939, when Germany was poised to invade Poland and start a world war, was not allowed to slow the sales machine. On 4 July 1939 Stanley sent each of his representatives a front page of that day's *Daily Express* with a story under the headline, 'Poles warn Berlin: Danzig arming must stop' next to an article about falling unemployment and the 12 810 000 Britons with jobs. C.O. thought this proved there was no reason for sales to be difficult, and explained that if they seemed hard to make it was because 'manufacturers have the jitters and we pass it on to the dealer'. The crisis in Europe was a chance to sell more radios. The trick was to 'inspire the dealer to divert the interests of the public to the need for buying better radio – radio that will get more distant stations – in these stirring times . . . Feel these things', he exhorted his salesmen, 'and you can make your dealers feel them'.

Someone who argued, and perhaps believed, that Nazi Germany on the move presented radio salesmen with an opportunity was not an easy man to put down, as Robinson now discovered. The two men certainly quarrelled over the making of radios, Robinson complaining to W.G. Pye's son Harold that Stanley's obsession with saving money was damaging the quality of Pye radios. Robinson was still officially in

Figure 3.20   Front cover of a leaflet

Figure 3.21   Finishing the assembly of Pye radiograms, 1936

charge of Pye's factory and laboratories, and any disagreements with
Stanley and Ellis, his fellow joint managing directors, were supposed
to be referred to the board, whose other members were now Polson and
the company secretary L.U. Slater. The stage was set for the only public
airing of doubts about the way C.O. Stanley ran a business until his fall
from grace 30 years later.

Over the next two years Stanley began to diversify Pye by setting
up or buying subsidiary companies. Maximum profit was to be taken
from the rental market by creating Pye's own firm, United Rentals,
which he argued would indirectly boost Pye sales; Robinson, though a
director of the new company, doubted dealers would find the idea attractive. Milward Ellis was authorised to set up a small subsidiary called
Cathodeon to carry out research into the cathode ray tubes used in the
latest electronic marvel, the television set. In fact, Cathodeon was part
of Pye, and separate only in its name and accounting. In November 1935
Stanley proposed that Pye 'acquire from him' Orr Radio and United
Radio, the kit-makers linked to the *Radio for the Million* campaign.
They were to be renamed as Invicta to sell to the cheaper end of the
market without damaging the prestige of the Pye trademark, and to pick
up business from the wholesalers Pye had abandoned.

In the spring of 1936 Stanley told the board that after looking at the
Irish market he had come to the conclusion that the only way to sell

Figure 3.22

in the Irish Free State, as Ireland was then called, was through a local company. He proposed selling the Pye trademark to this company, to be called Pye (Ireland), for £5000. Pye of Cambridge would invest a similar amount, representing 10 per cent of the new firm's authorised capital. In 1935 Pye began to make radios for the battery maker Ever Ready and its subsidiary Lissen. Ever Ready hoped this would increase sales of the batteries that remained its chief product. Pye made the chassis and cabinets for the sets and supervised their assembly at Ever Ready's London factory. Stanley claimed this would allow Cambridge to keep a more even workload throughout the year, but he also hoped to acquire Ever Ready valves on favourable terms, for he was always looking to buy at prices lower than those set by the British Valve-Makers Association.

Robinson was charged with conducting negotiations with Ever Ready, and the Pye minutes recorded no word of disagreement with Stanley over cost-cutting or anything else. Indeed the minutes made

Robinson seem the board's dominant member, and it must have looked from the outside as though the two men were working well together. Robinson thought so until February 1936, when a letter from C.O. hit him 'like a bolt from the blue'. For reasons that are not clear, Robinson was under the impression that they had almost reached agreement on a plan under which Stanley would gradually free himself from 'active participation in the managing of the radio group, retaining only an interest in general policy'. Stanley's letter made plain this was not his intention at all, but Robinson, in apparent desperation, now suggested the 'simplest way out' was for him 'somehow' to raise enough money to buy C.O. out. His ideas, he admitted, were 'confused'; all he knew for certain was that he could not go back to the days when the Pye leadership was divided.

C.O. must either have changed his mind and gone back on a deal that was, or so Robinson thought, almost done; or he had misled him from the start and had no intention of accepting just an 'interest in general policy'. In either case T.A.W. Robinson's days at Pye were numbered. When the managing directors' contracts came up for discussion the following month Robinson said he had reached a 'personal agreement with Mr Stanley', while C.O. proposed the board should 'compliment' Robinson, and also demonstrate to the outside world his continuing close interest in the company, by making him Pye's chairman, Polson (Stanley claimed) having agreed to retire in Robinson's favour. He also proposed giving Robinson a 3-year consultancy agreement at £1000 a year. His duties as joint managing director would be taken over by Stanley and Ellis.

It was an expedient formula. Stanley never meant to restore Ellis to power, but he needed him temporarily to strengthen his hand against his enemy of the moment. Robinson found himself without a real job and nothing more was heard of making him chairman. C.O. took over management of the Cambridge works. Polson bore no evident grudge against C.O. for suggesting he might be replaced, and was soon praising the 'remarkable zeal and ability with which he is endowed in the most inspiring way'. Robinson tried to go back on the deal, explaining that he wanted to be sure the company would be properly managed in his absence, but when Polson refused to cancel clauses in his consultancy contract that barred him working for a rival manufacturer he vowed to attend no more meetings while Sir Thomas was in the chair.

Matters came to a head the following spring when Robinson reappeared at board meetings after leave of absence in America and made an all-out attack on C.O.'s management of Pye. He queried accounts prepared for the growing number of subsidiaries, asked why Pye had agreed to pay another company's debt to C.O.'s own firm Arks, and how much money Pye (Ireland) owed to Cambridge. He did not like the way profits from subsidiaries were received as management fees rather than

Figure 3.23   The Pye factory, Cambridge, c.1937

Figure 3.24   Machine shop at the Cambridge factory, c.1937

Figure 3.25   Advertisement, 1937

as dividends, and questioned Pye's purchase of United Radio Manufacturers and Orr Radio from which both C.O. and Pearl benefited
as shareholders in the two companies. Robinson also demanded that
Pye directors be given more information about the accounts of subsidiaries, and that the minutes should record all authorisation of capital

expenditure. When he asked for the minutes of two previous board meetings to be circulated among all directors Stanley agreed, but said 'he was going to propose that the circulating of minutes be stopped and that the directors [only] be furnished with such extracts as they might require from time to time'. Robinson walked out of the next board meeting, and resigned his directorship in October 1937. The directors voted him a leaving present of a silver tea service. The next year he asked Pye to extend his consultancy agreement, and was refused.

Robinson seems never to have understood that once Stanley decided to be involved in Pye he would run it as he, and he alone, thought best. After Robinson's departure the minutes contained no more criticism of C.O.'s methods, for he was now getting the board he wanted. Leonard Tregoning, an able general manager brought into the Cambridge works at the end of 1934, possibly to undermine Robinson, became a director two years later. In 1938 Stanley's old associate and original Pye director L.G. Hawkins rejoined the board after Pye bought his electrical appliance business, to be followed by Stanley's sister Pearl and the radio engineer Charles Harmer. The two other directors, Ellis and Polson, had already shown themselves to be satisfactorily bullyable. Ellis did question the need to bring back Hawkins, but was easily overruled, while C.O. thought nothing of publicly criticising Sir Thomas if he made mistakes in his conduct of board meetings. The most independent-minded of the lot was Pearl, who was already being publicised in Invicta's advertisements as that rare creature, a woman managing director. She still referred to C.O. as 'my younger brother', and spoke out when she thought him wrong. There is no evidence, though, of her sharing T.A.W. Robinson's doubts about C.O.'s juggling of the affairs of Pye and its subsidiaries, or his determination to act alone and out of the view of others.

In 1932 Pye appointed a 26-year-old Hungarian inventor called Peter Goldmark to set up a television department. Goldmark took out Austria's first television patent when still a student in Vienna, and on the strength of it was invited to London by John Logie Baird, who was developing television based on mechanical scanning. When Baird would not give him a job Goldmark wrote to eight British radio companies and the only answer he got was from Milward Ellis at Pye. Once at Cambridge the young inventor switched to the more promising television system under development at EMI using a cathode ray tube and electronic scanning camera. Goldmark demonstrated a version of this to Prince George, later Duke of Kent, when he visited the Pye factory in 1932. Pye did its best to inspire their guest with the romance of new technology, the *Times* reporting that the royal visitor 'spoke into a microphone and observed the wave form produced'. Television,

Judge

for

yourself

People often write to me for advice about buying a new set. Frankly, I am inclined to tell them that our sets are as good as any that can be bought, but it seems to me that they would realise even before they wrote to me that I naturally have a high opinion of my own products. So I simply give them the name of a really dependable dealer in their district who stocks Invicta and three or four other makes of reliable receivers. I know how the man I recommend runs his business so they can be quite sure of a fair demonstration and judge for them-

selves which set meets their needs. I keep a record of these enquiries and it shows that more often than not the writer finally chooses an Invicta. This doesn't necessarily mean that our sets are better than any other on the market, but it does prove that letting a set speak for itself is the best advice that a manufacturer of any well-made receiver can give.

MODEL 430. *4-valve, 4-waveband Superhet for A.C. mains. Special Trawler, Police and A.R.P. wavebands. Spin-wheel tuning; tone compensated volume control; 4-colour edge-lighted tuning dial.* **£9.9.0**

Invicta sets are used by the Admiralty Fishery Protection Vessels, the Board of Trade and English Lighthouses. This type of receiver sets the high standard of reliability at which we have always aimed.

L. A. Shaw.

● *A comprehensive range of battery and mains Receivers and Radiogramophones, ...   from 5 gns to 50 gns.*

INVICTA
R A D I O

London Office: 44 Great Marlborough Street, London, W.1

Figure 3.26   Advert for Invicta, a subsidiary of Pye, 1938. Note the photograph of C.O. Stanley's sister, Pearl, known as Miss Shaw for business purposes

though, failed to fire the royal imagination. After inspecting Goldmark's prototype the Prince observed only that 'it would not replace cricket'.

Goldmark left Cambridge after 18 months when Pye told him, in words that might have been used by the young prince, that television 'probably would never be useful for the home'. (Goldmark later became head of the CBS laboratories in America and inventor of the long-playing record.) Within a year Pye had to reopen its television department as a matter of 'extreme urgency' after recognising that television sales would affect sales of radios. The board asked Milward Ellis to hire a new television team but it was Stanley, who had seen one of Baird's earliest demonstrations of his mechanical system, who became the new medium's prophet. Pye recruited Baden John Edwards and Donald Jackson, both formerly engineers with Standard Telephone and Cables (STC), the British arm of the American giant ITT. The pair were first interviewed by Ellis, Robinson and Tregoning, and it was only on a later visit to Cambridge that they met Stanley. Jackson described the encounter:

> He didn't ask us a single thing about engineering or technical things. He walked us around the factory, waving his hands up and down . . . telling us what he thought and wanted. He seemed more interested in the fact that Edwards and myself were both mad about cars and aeroplanes. We got on, and something must have clicked because Pye's factory was . . . very old-fashioned [by comparison with the modern laboratories at STC].

The two young engineers fell for Stanley because he shared his dreams with them. He persuaded them he could turn an indifferently equipped, middling manufacturer of run-of-the-mill radios into, as Jackson remembered him saying, 'one of the greatest radio and television companies in the world', and that the greatest satisfaction of their life would be to help him do it. It was a fateful meeting. There is nothing surprising about Stanley's confidence in the future television market. What was new was this demonstration of an ability to inspire the scientists on whose ingenuity conquest of that market depended. There was an echo here of the enjoyment with which he worked at Arks with creative but difficult talents such as William Connor and Donald Zec. B.J. Edwards' colleagues soon recognised him as a 'prodigy', but he was temperamental, over-bearing and often foul-mouthed. Donald Jackson was so obviously eccentric – he once took off the internal doors in his house so he could move around quicker – that he became known as 'mad' Jackson. Stanley ignored such things; what drew him to them and others like them was originality and daring, the qualities that were making him a successful salesman and entrepreneur.

Pye's decision to take up television again followed significant developments outside the company. A somewhat doubtful BBC had agreed

Figure 3.27   HRH Prince George, later Duke of Kent, viewing a Pye radio at the Cambridge factory, 1932

Figure 3.28   HRH Prince George with members of Pye Radio Ltd. staff at Cambridge, 1932. Note C.O. Stanley is second to the left of Prince George

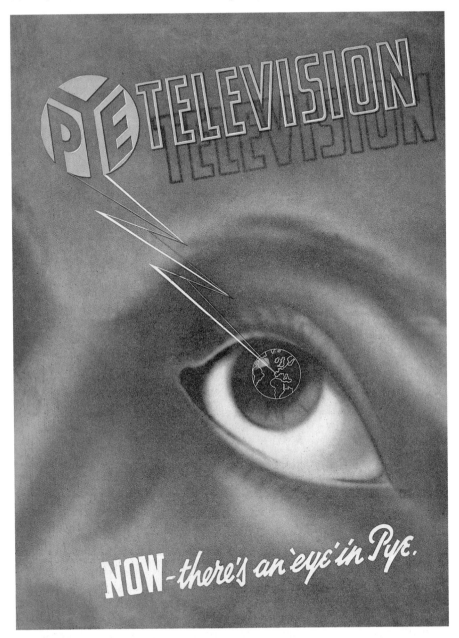

Figure 3.29   Front cover of a leaflet, 1936

with Baird in 1932 to provide two television programmes a week even though only 500 television sets were in use in the entire country. In 1936 the BBC began the world's first regular television broadcasts of just 2 hours a day from the Alexandra Palace in London. For a trial

Figure 3.30   Pye television receiver, model 4200, 1936. It also contained a radiogram with automatic record changer

period these alternated between the EMI and Baird systems, and sets had to be able to receive the broadcasts of both. Early in 1937 the BBC dropped Baird's technology using a 240-line picture in favour of EMI's clearly superior 405 lines (picture definition is largely determined by the lines: the greater the number, the sharper the picture on the screen).

This cleared the way for the new medium, even though television sets were still few and expensive. Short broadcasting hours and the cost of increasing their length and improving their content suggested expansion would be slow, yet it was plain that public interest was high from visitors to the annual Radiolympia where Pye and half a dozen companies first showed television in 1936. Pye signalled its change of direction by changing its name from the restrictive Pye Radio Ltd. to the all-embracing Pye Ltd. It was a moment that might have been contrived to excite C.O. Stanley, for it spoke to all his talents and convictions.

Figure 3.31   Pye television receiver chassis, model 4045, 1936. Note that the TV picture is reflected into a mirror from a vertical cathode ray tube

Here was a chance to market a new technology, to champion the mission of radio and television to entertain, and to attack monopoly control over broadcasting by the government, the BBC or any other establishment of the great and powerful.

At the birth of sound broadcasting the BBC's first director of programmes, Arthur Burrows, identified the new medium's challenge: if the BBC and its artists were being paid to broadcast they would have to adapt to an audience 'with a low proportion of persons who habitually attend the Albert, Queen's and Wigmore Halls'. In other words, they would have to learn to be popular. Surveys throughout the 1930s showed that Britain's less well-off radio listeners preferred commercial stations, the most popular of which, Radio Luxembourg, had 45 per cent of the country's Sunday listeners. The audience for the BBC seldom went above 35 per cent.

Stanley quickly saw the link between programme content and television sales, and managed to become chairman both of a committee appointed by the Radio Manufacturers' Association to promote television and also of its Television Development Sub-Committee, which sent fortnightly critical reports on TV programmes to both the government and the BBC. In his *History of British Broadcasting*, Asa Briggs calls the Stanley Committee's tastes 'undisguisedly lowbrow', hostile to ballet, opera and, as one report put it, 'morbid, sordid and horrific plays'. The government also put pressure on programme makers, warning that unless more people bought sets and took up licences (the fee was raised by 15 per cent in 1938) it would no longer pay the cost of television transmission. By the summer of 1938 Stanley was telling Pye shareholders that a 'marvellous improvement' in programmes had allowed television to grow faster than expected; the next year he was attacking the government for failing to extend transmissions outside London fast enough, and for ignoring the export potential of the 'prestige England has gained as the leader over the world in this new scientific invention'.

The combination of better programmes and cheaper sets raised sales in the last 3 months of 1938 to 5000, compared with only 2000 in the previous 9. Sales were also helped by the compactness of the new sets, made possible by smaller cathode ray tubes (Donald Jackson called the clumsy model Pye was working on when he arrived at Cambridge 'a bedstead'). The average factory price of the Pye sets designed by Edwards and his team had come down to £34, and the company's best

Figure 3.32   C.O. Stanley, in his capacity as chairman of the Television Development Committee of the Radio Manufacturer's Association, being interviewed on television about their plans for Radiolympia, 1939

Figure 3.33   Front cover of a leaflet, 1938

seller was a 9 inch model selling for just £30. Fearing prices might fall too fast, Stanley called a meeting of manufacturers in early 1939 to fix them, though he fought off suggestions that prices be raised.

By August 1939 the BBC was providing those lucky enough to live in London with the best television service in the world; 18 999 sets had been sold, more than in any other country including the USA. Pye sets were now the third best seller after EMI and Ekco (selling perhaps 2500 compared with Ekco's 3500 and EMI's 6000), largely thanks to the engineers Stanley had inspired to join him. The team's first results were shown at the 1936 Radio Show, though by then they had only completed a dozen sets. Their rapid progress did not make the company popular among its rivals. Engineers at Cambridge felt the bigger firms looked on Pye as 'a brash outsider [and] too pushy by half'. But when Pye displayed its 915 model at the 1939 Radio Show it could claim with some justice that the world's best television team had produced the best television set in the world.

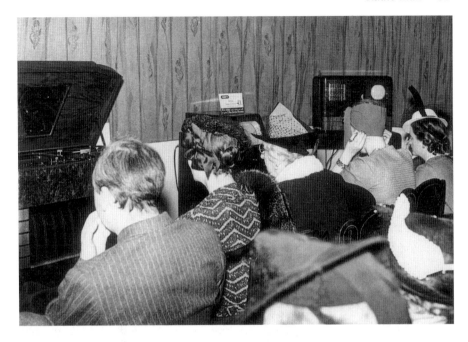

*Figure 3.34   At a Derby television party, 1939. Left to right: Compton Macken-
zie, Miss Fordham, Princesses Helena and Victoria, and Lady Iris
Mountbatten*

Not everything had gone smoothly. Both Cathodeon and another Pye
subsidiary Hi-Vac were unable to meet B.J. Edwards' specifications for
a new high-performance valve for the 915 set. The obvious solution was
to discuss the problem with Mullard, for the Captain and his engineers
were often at Cambridge, while Pye teams went regularly to Mullard in
London. The custom was that C.O. and Edwards went off with Captain
Mullard while the two companies' engineers discussed components and
future development work. Pye's Bill Pannell took part in these meetings
and was used to seeing Edwards turn up, give some directions and then
leave, for he was quickly bored and would not hang around. Pannell
decided that Edwards and C.O. Stanley had one thing in common – they
'didn't think like normal people' – and it 'sometimes [took] the rest of
us the day to understand what Edwards was getting at'.

This was how Mullard took up Edwards' parameters for the valve
known as the EF50. Eventually manufactured by Philips in Eindhoven
but stamped with Mullard's name it was the first all-glass valve without
a separate base to carry contact pins, but it failed to perform to Edwards'
expectations. With the August Radio Show approaching Stanley told
him to call in the best electronics engineers at the university to help solve
the problem, but Edwards and Jackson, working with the Mullard team,

Figure 3.35   Front cover of a leaflet, 1939

Figure 3.36   Part of a Pye 1939 television chassis, model 915, showing the EF50 valves that were subsequently used in many pieces of vital wartime equipment

came up with a mechanical solution of their own. First they improved the valve's performance by altering its base, only to get more distortion because of poor screening of the valve's inner surface. With no time to send valves back to Eindhoven to be rescreened, Jackson devised the 'Jackson cap', a small red spun-metal jacket placed on the top of the valve. Arranged after hours of experiment in an unusual sequence of five, it was largely the EF50 that allowed Jackson to call the 915 set 'the greatest [television] receiver in the world'.

The approach of war closed the Radio Show early and at noon on 1 September 1939 the BBC suspended television broadcasts. Not a single 915 set could be sold, but the ingenuity and sweat that produced it were not wasted. The configuration of five red-capped valves, soon to be known as the Pye strip, was to bring the company into the heart of Britain's secret war, and ensure C.O. Stanley a role in Britain's approaching fight for survival.

Sources

COS files: 2/17, 4/5 (annual reports).
COS/VDS1/1 (letters to Velma, courtship and struggle to control Pye).

Pye main board minutes.
Interviews: Dennis Fuller, Donald Jackson, Fred Keys, Jim Langford.
A–Z files: Fritz Philips (Stanley and Philips).
Books: Bell, *ibid.*; Bussey, *Wireless: The Crucial Decade*; Briggs, *The History of Broadcasting in the United Kingdom*, Vols I and II; Geddes and Bussey, *ibid.*; Goldmark, *Maverick Inventor*.

Chapter 4

C.O. goes to war

On 30 October 1933 C.O. wrote to Velma about some unusual visitors to the Cambridge factory. A team from the War Office had been to the plant and asked if Pye was ready to do more military work, 'or more particularly how far and how quick could we go in an emergency!' The exclamation mark suggests pleasure as much as surprise. In World War I W.G. Pye had met orders for a wide range of military equipment, but since then the government had given it only minor contracts. In the spring of 1939, after Britain guaranteed to come to Poland's help if it were attacked by Germany, Pye was told that in the event of war it would be classified as an official 'laboratory', though there was no hint of anything so momentous when Sir Thomas Polson spoke to the annual general meeting in June. He did not doubt, he said, that the trouble in Europe would be 'cleared up without resort to war', and even if conflict came there was nothing to fear for the outcome was sure to be 'an England mightier yet in all those things which make for ... peace, happiness and general well-being'.

C.O. was less complacent, and as preparations for war became more obvious his impatience to be doing something was clear to those around him. Even Dennis Fuller, then only a junior member of Pye's research team, sensed his employer was 'maddened not to be in the thick of it', and by the spring of 1939 C.O. was asking acquaintances how to get a war-related job in Whitehall. At least one friend told him it was a bad idea, warning it would only work 'provided you were placed in a position of authority; but that is the rub and I should hate to see you where you would probably be stymied unless you had a free hand!' Far better to serve the country by staying at his factory where, if war came, he would have his 'hands full'. How full, and how soon, not even C.O. could have guessed.

Looking for the enemy

In that same last spring before the war Edward Appleton, Harold Pye's old mentor and now Jacksonian professor of physics at Cambridge, sent a message to E.G. Bowen, another former pupil, who was pioneering radar at the highly secret Telecommunications Research Establishment (TRE) at Bawdsey Manor, a large Victorian house on the Suffolk coast. Pye, Appleton told Bowen, 'had set up a [television] production line for 45 Mc/s TRF (Tuned Radio Frequency) chassis' for its latest television set and already produced some trial models. In his memoir *Radar Days* Bowen describes how he reacted to this information like a runner to the starter's gun:

> I went hot foot to Cambridge to see B. J. Edwards . . . and was rewarded with a remarkable sight . . . he had scores of TRF chassis of just the type we were looking for. These used a new valve with an octal [sic] base that had not yet appeared on the market. It was the EF50, a valve which was destined to play almost as important a part in the radar war as the magnetron [see p. 104] itself.

There is no hint that C.O. knew anything of radar until Bowen's visit. B.J. Edwards had very likely grasped some of the implications of new work on radio waves, but it was a closely kept secret that work had begun

Figure 4.1   Radar towers at Bawdsey, photographed after World War II

Figure 4.2   E.G. Bowen, Pye's principal Telecommunications Research Establishment contact on radar, c.1940

in 1937 on 20 coastal radar stations stretching from Orkney to the Isle of Wight. Metro-Vickers and Cossor, the makers of the equipment, were told neither what it was for nor of the other's involvement (Marconi would provide the transmitter antennae).

The scientists involved in radar knew it was not a British invention. Researchers in several countries understood how radio waves could be projected onto remote objects and their distance from an observer on the ground measured by timing the echo's return – the technique used by Appleton when in 1924 he detected the ionosphere 60 miles above the earth's surface. The British achievement was not just to see the military implications of the technique, but to commit the money and resources to put it so speedily into practice. Germany had began to rebuild its air force in 1933 and the following year the government set up a Committee for the Scientific Survey of Air Defence under Henry Tizard, the Rector of Imperial College. Tizard's Committee first questioned Robert Watson-Watt, the plump and talkative Superintendent of the National Physical Laboratory's Radio Department, about the use of radio waves as 'death rays' against enemy planes. This idea was quickly discarded and Tizard then asked Watson-Watt to investigate the practicability of using radio waves to give warning of, and locate, approaching enemy aircraft. The result was Watson-Watt's rapidly written paper, *Detection and Location of Aircraft by Radio Methods*, followed on 26 February 1935 by a demonstration in which a plane flying in the 50 m beam of the BBC's Daventry transmitter produced an echo in a receiver 8 miles away. Tizard estimated that within five years it would be possible to direct fighters to attack enemy planes 100 miles away, and also to use radar to increase the accuracy of the notoriously inaccurate anti-aircraft guns. If he was right, Britain would have found the weapon to destroy Germany's advantage in the air. For security reasons the Committee christened the precious new technology RDF, radio direction finding, the term used by Britain until the United States entered the war and the American word 'radar' was adopted.

Government scientists built the first prototype at Bawdsey, whose position on the coast made it perfect for testing a radar supposed to give early warning of German planes approaching from the sea. When he was ready to build a series of coastal stations known as Chain Home, Watson-Watt recommended Metro-Vickers and Cossor for the job because he knew their senior research staff. Whitehall did not rate highly the British radio industry's capacity for research, judging only the General Electric Company capable of research of a fundamental nature. Other firms thought suitable to undertake research were Metro-Vickers, STC (which Edwards and Jackson had left to join Pye), British Thomson-Houston and EMI. Pye, with Cossor, Murphy and Ekco, was rated as having only small, though skilful, research teams. These schoolmasterish judgements would later be echoed in histories of radar, which concentrated on the achievements of Whitehall and its scientists while barely mentioning the equipment's makers. Watson-Watt, who did try to give industry its due, ridiculed this 'soothing fiction of an optimal requirement stated by an all-wise staff, and unquestioningly satisfied by a docile and expert developer (ie company)'. The development of radar was seldom to be like that; at Pye under C.O. Stanley, never.

In spite of some investment by C.O. the Pye laboratories still looked as unimpressive as on the day Donald Jackson and B.J. Edwards first saw them. Anita Sturrock, joining the research team at the start of the war, was horrified by what she found. Scientists with university training such as the Cambridge graduates Dennis Fuller and Donald Weighton were the exception. Edwards and Jackson were polytechnic boys, but most of the 30-odd staff had learned the job as apprentices. Sturrock had accompanied her father, a structural engineer, on visits to British power stations and factories, and by comparison with what she saw there the Pye labs were 'a disgrace': dingy, poorly lit and ventilated, and lacking the necessary work tools. Chronic underinvestment, she thought, had left it ill-equipped to meet the demands of war.

Pye's facilities, or lack of them, were perhaps an extreme example of an industry that Corelli Barnett in *The Audit of War* characterised as 'prosperous, [but] in no sense a world leader'. In spite of C.O. Stanley's attempt to develop and make his own valves and cathode ray tubes Pye remained, like its competitors, largely an assembler of bought-in components and an employer of much unskilled and seasonal labour. Its world-beating 1939 television receiver was only possible because the Dutch giant Philips had been able to make the revolutionary EF50 valve for B.J. Edwards after he had failed to get it from Pye's own subsidiaries.

Bowen was overjoyed to discover the EF50 in Cambridge because Tizard, suspecting Germany might use night-bombing, had asked him to make a radar small enough to fit into a night fighter, or for use in planes attacking ships and surfaced submarines. No one knew if this was possible. The transmitter of the prototype Chain Home radar at Bawdsey weighed several tons and had an aerial 75 ft tall; Bowen's task was to build a set weighing 200 lbs with an antenna just 1 ft long. The Chain Home stations floodlit a wide area with their beams and lacked accuracy, but the ultimate aim of radar researchers was a narrow, pencil-like beam that could only be generated by using ever shorter wavelengths and aerials. This, it was calculated, would achieve greater accuracy of location, and cut out misleading echoes. The key to progress were valves with greater power of amplification that could function at ever higher frequencies.

Unable to find British components of sufficient quality Bowen used American short-wave transmitting and receiving valves in his first experiments. Before his dash to Cambridge, he had managed to make a radar that when installed in an RAF Anson picked up targets at almost the maximum required range of 3–4 miles, and with the minimum range of 1000 ft thought necessary for pilots to identify enemy planes visually on dark nights. Watson-Watt gave a preliminary order for six each of Bowen's transmitters and receivers to Metro-Vickers and Cossor, the companies already making the equipment for Chain Home. Cossor, using the EMI television chassis adapted by Bowen for his test model, quickly produced six receivers that a horrified Bowen judged 'a complete failure . . . [not] within a factor of ten of the required sensitivity . . . [and] weigh[ing] more than our entire system'.

It was shortly after this setback that Bowen went to the Pye factory, where he found other treasure besides the EF50. He also took away from that visit samples of the Pye 1939 television chassis which was smaller and lighter than EMI's, and also a Pye cathode ray tube that he judged well-suited to airborne radar. After making modifications of his own, Bowen thought he had a workable set. Orders were given to Metro-Vickers for 30 transmitters, and Pye, from which Whitehall had hitherto expected little, was asked to make 30 receivers plus most of the switch gear, power supplies, control panels and other components necessary for a complete assembly. From this moment C.O. and his company were caught up in the rush of a war whose demands were as unpredictable as they were fast-changing. This order was supposed to be finished and fitted in 30 planes within 30 days, a deadline that revealed London's now acute sense of urgency rather than a grasp of the possible. Bowen had a team of just 23 scientists and technicians, and such rapid work was only possible with the help of Pye's television engineers, who overnight became its radar team. They travelled between Cambridge, Bawdsey and its local airfield at RAF Martlesham Heath carrying parts and technical drawings, only breaking off to finish work on Pye's TV sets for the August 1939 Radio Show. C.O. was in the thick of it, and war had not even been declared.

Bowen found that the new receivers performed well, largely thanks to the improvement in sensitivity that the Pye receivers made possible. The sets were installed and tested at Martlesham in Blenheims standing on an open runway exposed to wind and rain. Several members of the Pye team, including B.J. Edwards, flew in the first trials – Edwards' and Jackson's experience as amateur pilots gave them a feel for the sort of equipment needed to withstand the stresses of flight and careless treatment by air crews. Blenheims had been chosen as the first plane to receive the sets because one of the two-man crew could operate the radar, but the planes' vibration damaged the receivers and so did altitude, and this meant more work to repair and service them. Co-operation between Pye and TRE was not always easy. Edwards cursed the government scientists for making impossible requests and asking stupid questions; an essential part of Pye's job, Anita Sturrock thought, was 'turning pie-in-the-sky ideas into something practicable'. Sir Charles Oatley, one of the first Cambridge scientists to be initiated into the radar secret, later acknowledged that only Britain's pre-war progress in television produced experienced engineers capable of taking the models designed by government scientists and 'engineer[ing] them for production to withstand the rough handling of the services'. The first radar-equipped Blenheim was delivered to 25 Squadron at RAF Northolt in mid-August and, by what Bowen called 'prodigious efforts', another five by the end of that month. It was a close-run thing. On the first night of the war just one radar-equipped Blenheim nightfighter was ready to patrol the sky over London.

It is not clear when C.O. understood that his new television set was of national importance, but it cannot have been long after Bowen's visit. Pye's sudden involvement with this most secret weapon at once brought down MI5 officers to investigate the factory and its staff, including C.O., whose Irish background gave the security men a moment's pause. They ordered that curtains made of blackout material be hung up to shield all work in the Upper Laboratory that was now devoted to radar, a useless precaution in the case of Donald Jackson, who could never keep quiet about what he was doing. Official secrecy, though, did ensure that Pye kept few records of its work on airborne or any other sort of radar. Throughout the war the Ministry of Information censored both the company minutes and Sir Thomas Polson's speeches at annual general meetings. Engineers who worked in the Upper Laboratory felt gagged by official secrecy all their lives and never published accounts of their achievements. C.O. himself was not one for records at the best of times: it was said at Pye that he only bothered to write down figures if they were very large, and then only on the back of an old envelope. And the conditions of war did not encourage careful note-taking. Apart from spells of working seven days a week at the factory, Pye's radar engineers were often on the move, if not to Bawdsey or an airbase then to London where many of the later discussions about design took place with TRE. There was nothing orderly about these meetings usually held sitting on hard chairs outside the office of an Air Ministry mandarin, or late at night in a hotel lounge after a long day's work somewhere else.

Chain Home, Britain's first working radar defence, had a weakness: it could detect neither planes that flew low towards the coast nor ships and submarines approaching on the surface of the sea. It was also discovered that while Chain Home could direct British fighters accurately in the direction of oncoming German formations it could not be relied on to put nightfighters into their best attacking position above the target. The solution was to add a supplementary net of radars using the shorter wavelengths of Bowen's infant airborne sets. The decision to improve on Chain Home at once brought more work for Pye. In August 1939 a team of 80 physicists led by John Cockcroft, Jacksonian professor of natural philosophy at Cambridge, was preparing to spend a month learning radar techniques. In early September, though, Watson-Watt instructed Cockroft, who was soon to become the Army's radar supremo and later a Nobel prize laureate, to build three Coastal Defence stations on Fair Isle and Shetland capable of detecting surfaced U-boats. Cockroft began work on this equipment in the High Voltage Laboratory at Cambridge. The guts of his receiver was the same Pye TV used

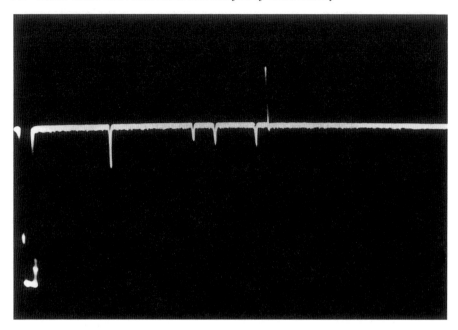

Figure 4.3   Cathode ray tube display of CH (Chain Home) ground station. Aircraft appear as V-shaped echoes below the horizontal trace. Distance from the large transmitting pulse on the left gives range measurement

by Bowen, and a team from Pye found itself working alongside the Cavendish scientists.

The Scottish stations were still under construction when war broke out and Germany's own secret weapon, the magnetic mine, began to litter the east coast with the wrecks of British merchantmen. The German planes that dropped the mines flew too low to be detected by Chain Home, and Watson-Watt asked Cockroft to turn Coastal Defence radar into a system called Chain Home Low (CHL) with the urgent aim of stopping German mines closing the Thames Estuary to British shipping. A party including Cockroft and a young Air Force officer Ian Orr-Ewing (later a Tory MP and life peer) went to Cambridge in mid-November 1939 to see what Pye was capable of, and returned in triumph to London with samples of the EF50. Though the valve was already known to Bowen's little airborne group and Cockcroft's crew in Scotland the secrecy surrounding radar projects was so great that knowledge was shared neither among government scientists nor military planners.

Bawdsey's original prototype for CHL had not used the EF50, and once TRE scientists had again established its better amplification when configured as the Pye strip, Cockroft asked C.O. to make the receivers and aerials for the new radar. A pilot station went on air on 1 December,

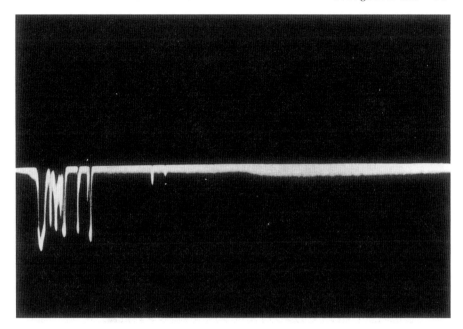

Figure 4.4   Cathode ray tube (CRT) display of CHL (Chain Home Low) which gives coverage for lower flying aircraft than CH (Chain Home) equipment does. When the radar beam illuminates any aircraft a response appears on the CRT at the appropriate range, which is measured from the left-hand transmitting pulse. The two V-shaped echoes one third of the way along the trace are from single aircraft at 50 and 55 miles

and that day an Air Force controller for the first time managed to direct British fighters towards low-flying German planes as they were laying mines. The new German threat was judged so serious that in January 1940 Cockroft and Watson-Watt summoned B.J. Edwards to a meeting in London that called for 'panic measures' to allow speedy completion of 7 of the 25 CHL receivers Pye had begun making (a later meeting recognised Pye's need for better machine tools if it was to build sophisticated equipment at such speed). Out of this advance came the idea for Ground Control Interception radar (GCI), the heart of Britain's future night defence against German bombers.

Cockroft was not universally popular in Whitehall, where he was known as 'the comet' because he moved fast and left a trail of wounded bureaucrats behind him, but these were qualities that endeared him to C.O. His engineers liked Cockroft too (the sometimes self-important Watson-Watt went down less well with them), and whenever C.O. brought him to the Pye laboratories the required work was done with particular care and speed. The physicist had gone away from the mid-November meeting thinking Pye had a large stock of the vital EF50 valves. It was true C.O. had a warehouse at Newmarket full of them, and that they only needed to have their red caps painted grey to make them ready for military use. 'We did indeed have "plenty" [of EF50s]', Dennis Fuller recalled, 'but "plenty" meant in terms of our requirements to launch a TV set'. C.O. was the first to see that operational radar sets would need many more valves than his store-rooms held and on his own initiative contacted the valve's maker, Philips. According to Fuller, when C.O. discovered that the Dutch had further large stocks in Eindhoven he turned into

> a demon. [He was] desperate to convince the government that they must send a ship to collect the entire stocks of the EF50 as soon as possible. They wouldn't listen, or if they did they didn't seem able to find a ship . . . [but C.O.] wouldn't stop. He rung and pestered everybody he could think of . . . [and] then started sending people in the ministries letters and telegrams. He finally told some general . . . that if they didn't collect the valves immediately he personally would send two lorries to Harwich, . . . hire a boat and . . . go and collect the valves himself.

The government eventually recognised the emergency, but only in the nick of time. In March 1940 it invited Dr Th. P. Tromp, general manager of Philips Electronic Valves, to London, where at a secret meeting Watson-Watt asked him to send to Britain all the EF50 valves and the tools to make them. Tromp loaded up a Dutch merchantman which evaded a German air attack to arrive at Harwich on 9 May with 25 000 valves and 250 000 valve bases. Hours later the German army invaded the Netherlands. The precise extent of C.O.'s contribution to this coup is difficult to establish, but that he was the first to spot the problem, and that he then harassed, and in many cases made enemies of, dozens of people with what Fuller called his 'bloody-minded' persistence there is no doubt. That was the way he went to war.

The development of CD/CHL brought an even heavier workload to the Cambridge factory. Soon the order for CHL stations rose to 40, of which 25 were to be mobile. Pye undertook to provide the receiver rack and cathode ray tubes, with the Pye strip fitted into a receiver designed by TRE's W.A.S. Butement. So close was the collaboration between the two teams that it was quite usual for a Pye engineer to telephone Cambridge from a CHL base and read over necessary modifications to the receivers as they were dictated to him by a member of TRE. The mobile CHL stations were mounted on two trucks for which Pye made antennae out of chicken wire and Jackson designed turntables and steering gear. That work, at least, was a poorly kept secret for he tested his early models on the factory roof, and once left his designs for them in a restaurant where he had stopped for a meal after a visit to TRE. Some

Pye engineers spent up to two years working on CHL, first travelling the country to erect new stations, later doing the unglamorous, and little remembered, job of servicing them. When the Ministry of Supply placed orders on the Army's behalf for another 120 CHL stations in the summer of 1940 it stressed that the highly specialised receivers and antennae could only be made by Pye, and that it was 'essential for control to remain in [Pye's] hands'. C.O. brought in his subsidiary Invicta to help with this order, and together they reached a production rate of 24 receivers a week, even though Pye was at the same time meeting ever growing orders for sets for the Air Ministry.

The factory now had so much to do that the development team might work for days on end on an urgent project, and sleep in the labs at night. C.O.'s management style, as described by Donald Jackson, contributed to the pressure.

> [He was] a great one for saying [to Cockroft or some other important visitor] "Leave that to us. We'll find a way, we'll sort that out for you." Then there'd be a big fuss because existing work had to be re-allocated. Everything ... was urgent but we kept on taking more and more jobs to help this or that development. ... Some people in the factory complained the place was chaos, and we shouldn't be taking on things that we didn't have the people to deal with ... properly. There were times ... when we got ourselves into some terrible muddles, taking on too many jobs at once.

From 1939 to 1941 Pye worked on two sorts of airborne radar, Air-Interception (AI) for night fighters and Anti-Surface Vessel (ASV) to detect ships and surfaced submarines. Bowen calculated that when the first 30 AI sets (AI Mark I) were completed Pye at once started work on 300 Mark II models; in 1940, in conjunction with EMI, it made 3000 Mark III and IV sets, and a further 3000 of the Mark IV in 1941. Each model represented an advance, and progress was so fast that equipment might be out of date by the time it was installed in aircraft. Co-operation between TRE and the manufacturers was so intense that the Air Ministry posted a liaison officer to Cambridge, while from the end of 1939 Bowen's second-in-command, Dr A.G. Touch, was spending most of his time conferring with Ekco and Pye.

At the start of 1940 few people even in the RAF had heard of AI, but the opening stages of the Battle of Britain in June and July brought home its importance. On the night of 22/23 July a radar-equipped Blenheim made the first night-time kill of a German plane, a Dornier. The Blenheim belonged to a special unit set up by the RAF to work out the best way to use AI. Most air crews were still untrained, or at best inexpert, in handling the new equipment, and this and the Blenheim's own inadequacies meant German planes still had relatively easy access to Britain's cities at night. The situation improved with the entry into

Figure 4.5   At the Telecommunications Research Establishment. Politicians, senior scientists and aircrew gather for one of Dr A.P. Rowe's informal meetings where problems were thrashed out without regard for seniority or rank amongst the people present (Sir Robert Renwick is sitting 4th from the left)

Figure 4.6   AI (Air Interception) indicator and receiver installed in a Mosquito night fighter

service of the faster and better armed Beaufighter, which also had more space for AI equipment, and the introduction of Ground Control Interception (GCI) radar that put fighters accurately on the tail of German planes. Pye built the receiving equipment and aerials for the new GCI as it had for Chain Home, the work being supervised by Leslie Germany, a member of the pre-war television team who became C.O.'s specialist in ground control radar. For a time he had his own RAF-manned GCI station at Sawston just outside Cambridge where he tested new aerials, anti-jamming devices and other novelties before installing them throughout the GCI net.

When London came under air attack Leslie Germany's team used GCI technology to build a gun operation room in the old Brompton Road tube station, and put its receiving aerial on the top of Tizard's Imperial College. This centre received information from the circle of GCI stations around the capital and fed it into a central plotting room. Here it was displayed on a projection tube specially developed by Cathodeon to hold the blips made by enemy planes for up to 30 seconds and back-project them onto a circular plotting table. Leslie Germany was then called to RAE Farnborough to work on a combined tracking station nicknamed 'the Happydrome' after a wartime radio programme. This sophisticated system used 13 consoles to give a composite picture of the location and direction of both German and British planes, and the Pye team spent many months completing a trial station near Christchurch. GCI proved its worth. During heavy raids in March 1941, RAF night-fighters brought down only 22 German planes. By May improved radar control allowed them to destroy more than 100. Germany could not sustain such a rate of loss, and from June the raids tapered off.

If the EF50 put C.O. at the heart of the radar war, he used all his talents to make sure that he and his company stayed there, setting out to charm the important visitors who now often came to Cambridge. Pye did relatively little work for the Royal Navy. Its range-finding equipment contributed to the destruction of the German battleship *Scharnhorst* in 1942 and it made receivers for submarine-detecting Asdic, but in general C.O. thought Admiralty orders too small for efficient production runs. That did not stop him being 'exceptionally helpful' to Professor John Coales, one of the Navy's chief radar scientists, whenever he visited Pye. C.O.'s knack of making himself helpful had much to do with the way he ran the company. Brigadier E.J.H. Moppett, an official in the wartime Ministry of Supply, told C.O. that what made Pye different was its

> fantastic speed of conception and realisation ... Your ruthless cutting of red tape and of the old slow processes created an environment and a spirit found far less at other firms (and ... even less at the government establishments). [This] actively generated new ideas to meet new situations and carried them rapidly through to practical form.

Not everyone found C.O.'s style so praiseworthy. Confidence in Pye's ability reinforced his natural self-assurance to a point where he insisted on his right to decide how new radar equipment should be made. A Ministry of Supply team visiting Cambridge in August 1940 to discuss orders for the Army's CD/CHL receivers thought Pye too cockily confident that their receiver design was better than the Army's. An official noted that Pye took a 'very high-handed attitude', behaving as though 'their pre-war television experience gives them the right to lay down the law'. If it was good enough for the Air Ministry and use in planes, C.O.'s engineers said, 'it must be good enough for the Army'. In notes written for fellow directors at the end of 1943 C.O. would boast that Pye always refused 'any instructions that we did not think in the interests of the country'.

Pye continued radar work with ASV sets for the war against German submarines, making, in co-operation with Ekco, 300 ASV Mark I sets in 1939, and a total of 6000 Mark II sets from 1940 to 1941. The Mark II, the first properly engineered ASV set, saw more operational service than any other radar of the war. Its greatest hour came in late

1940, when German submarines in the Atlantic threatened to isolate the British Isles. ASV allowed British planes to spot U-boats when they came to the surface to recharge batteries or to travel at maximum speed; forced underwater by the threat of detection by ASV they lost mobility and could do less damage. By the end of 1941 the improved ASV Mark II helped raise German losses to the unsustainable rate of one submarine for every ten Allied vessels destroyed. Radar had redressed another deadly imbalance.

Radar development had to be nimble to follow the war's changing patterns, and C.O. enhanced Pye's nimbleness by recruiting both undergraduates and graduates from the university. An important new pattern was Britain's switch from defence in the air to offence, and in 1941 TRE set out to improve bombing accuracy with Oboe, a system of pulse signals transmitted by two ground stations to time the accurate release of a plane's bombs over target. In early 1943 fast-flying Mosquitoes fitted with Oboe led the raids on the Ruhr, where much of German industry was concentrated, marking with firebombs targets for the heavy bombers that followed. As in so many radar projects Pye had to engineer the display and computing part of Oboe at such speed that there was no time for accuracy trials. Three months after Germany's surrender TRE told C.O. that bombers equipped with Oboe had achieved accuracies of 35 yards at 5000 ft, and of 100 yards even at 25 000 ft – very smart bombing by the standards of the time.

The approach of D-Day brought requests for other new equipment. Since 1939 John Cockroft had tried to use radar to enable searchlights to find and fix German planes. Early types of Elsie (Searchlight Control System) worked poorly, and even worse once German planes began to jam British radar signals. In 1944 Cockroft's team designed a centimetric version of Elsie (for centimetric radar see p. 104) and asked C.O. to make 24 of these SLC8(X) sets at the greatest possible speed. The Pye team that tested this novel device on the factory sports ground included C.O.'s son John, who had just finished school at Stowe and was about to read engineering at King's College, Cambridge. The first of the new model Elsies reached British troops in France a month after D-Day. Pye also rushed to meet the D-Day deadline for Watchdog, a system mounted on a half-ton truck that gave warning of oncoming vehicles.

As the quantity and variety of radar equipment increased it was discovered that poor matching between cables and connecting plugs often produced a 'reflection' that distorted the return signal, making it difficult to interpret. Donald Jackson solved the problem with what came to be known as the Pye plug, a coaxial plug that flexibly linked the cables between a radar set's various components. By the war's end factories in the United States, Australia and Canada as well as Britain had made more than 20 million Pye plugs, ancestors of the connecting plugs now found on every television set. The radar war also led to the adaptation

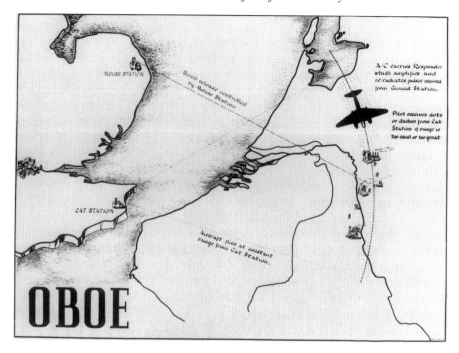

Figure 4.7   Diagram showing how the Ground Station in England controlled the release of bombs in raids over the Ruhr

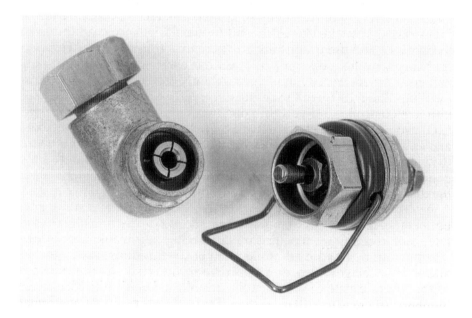

Figure 4.8   Donald Jackson's coaxial plug – known as the Pye Plug

of another Pye technology for military purposes. The radar-equipped Mosquito fighter-bomber had a wooden airframe for which its makers De Haviland constructed a gantry to hold it fast until its bonding glue dried. C.O. and his engineers were often at De Haviland installing radar receivers and they knew about this time-consuming process that made rapid production impossible. Inspired by Winston Churchill's call for more aircraft, C.O. set off for De Haviland with John Pound, a young engineer who had developed a technique called Radio Frequency (RF) heating that allowed Pye to speed up the gluing of its wooden radio cabinets. Pound showed the aeroplane makers how to halve the drying time by passing an RF down the glue line between the joints of the Mosquito's airframe.

The initiative over RF heating was typical of the behaviour that brought C.O. to the attention of Sir Robert Renwick, Controller of Communications at the Air Ministry, and also of Communications Equipment at the Ministry of Aircraft Production (MAP). Both Renwick's jobs were vital to the production of radar for the Air Force and, when offered the choice of either, he had asked to do both simultaneously so as not to have to ring 'the other fellow'. A background more different from C.O.'s was hard to imagine. An old Etonian baronet who, when war broke out, was Chairman of his family's London Electricity Supply Company, Renwick moved through the City, the grand London clubs and Whitehall with an ease C.O. could not hope to match. His twin positions in the Air Ministry and MAP made Renwick master of industrial support for the RAF, and he would bully, charm (he liked to be liked) and use every trick he knew to get things done. This was probably why he took to C.O. at a time when many in the services and government were turning against him. As chairman of the regular meetings that monitored the progress of radar production Renwick was ultimate controller of Pye's programmes, but there is no evidence that C.O. ever seriously quarrelled with him. That Renwick was exuberant and genial (he was widely known as Uncle Bob) goes some way towards explaining why C.O. took to him, and why Renwick was amused rather than irritated by the Irishman's quirks. What certainly brought them together was a shared passion for getting things done.
Well-prepared though Britain was for the radar war by the government's foresight over Chain Home and the industrial advances in television technology, the electronics industry as a whole was far less ready for war. It had made little attempt to standardise parts, and as late as 1943 the government's Radio Production Executive complained of the 'chaotic state of non-interchangeable and unidentifiable service numbers' (there were still more than 1000 different types of transformer lamination, 800 types of resistor, and 15 different magnetrons). It was Renwick's job to grapple with this and other production problems, and

to help him he recruited a team of over 100 engineers and accountants known as 'Renwick's Boys'. One of them, Gordon Maclagan, who had the task of organising the 130 companies that supplied electronics to MAP discovered that only 5 of these firms were using proper planning procedures when war broke out. Pye was not one of them, but C.O. quickly understood that MAP and its associated companies would only get money quickly from the Treasury if they showed efficient production planning, and he made an unexpected appearance at one of Maclagan's first monthly meetings with the industry. Maclagan remembered how C.O. stood up in a rather grand way and said, 'the work you are doing is vitally important and I shall do everything I can to support you'. Maclagan was pleased but also embarrassed, for most companies had sent only their production managers and 'not all the other [chiefs] were very pleased to be shamed into attending by C.O.'.

C.O. made a different impression at a meeting of a grander sort held weekly by Renwick and attended by Air Marshals, senior civil servants and, sometimes, business leaders. After listening to a civil servant talk for several minutes about the importance of keeping to 'correct channels' C.O. could not contain himself. 'Never mind whether the channels are correct', he interrupted. 'Are they effective?' He was not invited again. If Renwick decided to keep C.O. away from the top brass he valued him no less, and put him in charge of a committee set up to tackle the late supply of components that was playing havoc with MAP's production plans.

C.O. happily seconded senior staff to MAP and paid their wages, and even lent it the irreplaceable B.J. Edwards for seven months. There is reason to suppose that Robert Renwick shared the Pye view that C.O. was MAP's 'most dynamic and effective supporter' in industry. Pye's relative smallness was an advantage here. Brian Callick, a member of TRE, observed that while the biggest firms were often reluctant to try new ideas, smaller ones, and Pye in particular, 'had an instinct for innovation and improvisation', scarcely a surprise in Pye's case since these were among the strongest talents of both C.O. and B.J. Edwards. Maclagan came to rank C.O. with 'Ginger' Lee of the aerial makers Belling-Lee as MAP's 'unconditionally supportive industrial leaders', both having a 'gift for seeing the wood through the trees, and . . . content to set aside . . . the rivalries and jealousies that made some of the firms such a nuisance to deal with'. Certainly C.O. told his employees to co-operate in full with other companies over war work. 'Don't keep anything under the counter', he would say. 'Tell them everything. After the war credit will be given where it's due.'

C.O.'s liking for Renwick did not stop him defending his own interests. In the 1943 note to Pye directors he stressed he would only do development and preproduction work for MAP because frequent changes to its orders made smooth production runs impossible. If

Renwick knew about this he did not seem to mind. C.O.'s greatest contribution to MAP, Maclagan thought, was 'his readiness to "take a view" about what was important'. Robert Renwick was better placed than anyone to appreciate that, and it was the foundation of a friendship with unexpected importance for post-war Britain.

The deadly fuze

When the first German V2 rockets hit Britain in September 1944 B.J. Edwards was unimpressed. 'There's nothing you can tell me about V2s', he said to his research staff. 'Designed one of my own before the war. Quite straightforward really.' It was true. In 1939 Edwards built a 10 ft plywood model of a rocket outside the factory building, and experimented with putting television cameras in its nose cone. When war broke out he got Dennis Fuller and Pye's best mathematician Donald Weighton to experiment with wire-guided missiles. They made blanks with 2 miles of fine wire coiled up in the back of each shell, and tested them with Jackson's help on the War Office firing range at Shoeburyness. The Pye design was later passed to America, but nothing more was done about it then at Cambridge. Other exotic weapons were tried and abandoned. A young scientist called Jan Forman, briefly employed at Cambridge, invented a 'magnetic' anti-tank gun that fired a plastic missile filled with phosphorous that spread flames over anything it hit. General Sir Frederick Pile, Commander-in-Chief Anti-Aircraft Command, saw it when C.O. showed him round Pye in April 1941 and was sufficiently impressed to tell Cockroft about this 'electric gun' that was made to plug into an ordinary domestic power socket (when the German invasion did not come this handy weapon for citizens' defence was abandoned). Another forgotten device was an anti-submarine detector based on radar principles. Its inventors camped for months in a caravan parked by a Cambridgeshire gravel pit, submerging their device in the pit waters to see if it could find objects they dropped there.

But nothing was more exotic, yet in the end more effective, than Pye's work on the proximity fuze that would revolutionise anti-aircraft defence. The achievements of what came to be called the deadly fuze are still not widely known, largely because the British know-how was handed over to the United States, where final development and mass production of the fuzes eventually took place. This was a result of Anglo-American talks that began in Washington in September 1940, more than a year before Japan's attack on Pearl Harbor brought America openly into the war. The British side, led by Henry Tizard and including C.O.'s new scientist friends Cockroft and Bowen, took with them a black deed box and 12 crates containing papers, blueprints and samples representing Britain's most precious scientific secrets. Radar's recent past was represented by Pye-made ASV Mark II and AI Mark IV sets. Ted Cope from the Upper Laboratory accompanied them, but was not allowed to see where the Americans installed his sets. Radar's future was represented by the cavity, or resonant, magnetron. Since the beginning of the war British scientists had been trying to make valves that could generate sufficient power to operate on wavelengths of less than 1 m, the precondition for radars of greater accuracy and sensitivity. This goal was achieved not by a valve, but by the cavity magnetron developed by GEC and Birmingham University. For the Americans the magnetron was, in Bowen's words, a 'gift from the Gods', allowing them to apply their far greater scientific and industrial resources to the development of a new generation of centimetric radars with a performance far outstripping Pye's early airborne sets. Among these and other priceless items was the proximity fuze.

Tizard's talks were the most far-reaching exchange of military secrets ever to take place between two countries: the Office of Scientific Research and Development, which controlled America's wartime research, called the information Britain handed over 'the most valuable cargo ever brought to the shores of the United States'. The British government had hesitated before taking this unprecedented step, and many in Britain were never reconciled to it, C.O. among them. 'Scientifically and industrially', he wrote later with characteristic extravagance, 'the radar that won the war came from Cambridge, but it was a tragedy that we should have allowed the Americans to steal the trademark.' This only happened, he explained, because the threat of German attacks on British factories made it prudent to transfer much military production to America.

C.O. certainly felt that Pye was wronged. Had he not told his engineers to 'forget patents' and co-operate with everyone for the sake of winning the war because in the end they would get the credit they deserved? But he would never admit that the real reason for giving such treasures to the United States was Britain's inability to make full use of them on its own. The EF50, C.O.'s entry ticket to radar, was a Dutch-made valve that British brains had devised but could not manufacture. C.O.'s engineers knew better than anyone how poorly equipped their factory was for war. Their precision tools were almost all German-made, and the chronic shortage of both tools and components was only relieved when supplies from America began to arrive under the Lend-Lease programme. Pye survived and even thrived, the young engineer Anita Sturrock thought, on 'a genius for improvisation' and a willingness to be 'absurdly overworked'. Small wonder a meeting of British service and science chiefs was called in the spring of 1940 to discuss Britain's inadequate industrial capacity. Watson-Watt argued that the difficulties were temporary. There was no need to pass information on radar to the United States, he said, because by the end of the year

British production facilities would be better than America's. If that was so, Tizard remarked to assent from Cockroft, 'it would be the first time in history that this has happened'. The story of C.O. Stanley, Pye and the proximity fuze showed how right Tizard was: Britain gave America its inventions because, to win the war, it needed America to exploit them.

Tizard and Edward Appleton recommended that the government set up an Inter-Departmental Committee on proximity fuzes only three months after the outbreak of war. The idea for the fuze was born out of the unsatisfactory performance of many munitions, above all anti-aircraft shells. Few people in Britain knew that the anti-aircraft guns of 1939 gave little defence against German bombers. Jack Allen, a Cavendish Laboratory scientist who worked on the proximity fuze, calculated that during the London Blitz AA guns nightly fired off 30 000 rounds and wore out 20 gun barrels but brought down just 2 or 3 German planes. And since half the shells left the factory with faulty fuzes and exploded only when they fell back to the ground they might kill as many Londoners as the German bombs. Gun crews were supposed to set these traditional AA clockwork fuzes before firing to ensure explosion within lethal distance of attacking planes, but the necessary calculations of height, direction and speed, difficult at any time, were almost impossible in the stress of battle. It needed no great imagination to think of a smart shell that would know when it was within range of its target, and explode of its own accord. Similarly smart bombs, rockets and torpedoes would be useful too. The answer was to make a device small and hardy enough to fit into a shell and act as its brain.

The problem attracted scientists in Britain, America and Germany before the war began, and by 1940 it was also absorbing C.O.'s young inventor Jan Forman. When General Pile toured the Cambridge plant in April 1941 he saw something that intrigued him even more than the inventor's 'electric gun'. Forman, he wrote to C.O. after the visit, must be 'encouraged over the radio fuzes. It is a complicated mess of tricks, but if it does even a quarter of what seems likely, it might have revolutionary effects on Anti-Aircraft gunnery'.

Forman had begun work at the start of 1940 with the aim of putting a tiny transmitter and receiver inside a shell to set off the fuze within killing range of a plane. At first the Pye researchers were not sure what would trigger the fuze, for example an aeroplane's sound or its magnetic effect, but they knew they needed a valve small enough to fit in a shell's nose cone, and strong enough to withstand the shock of speed and spin when fired from the barrel of a gun. In the spring of 1940 Cockroft came across some small and interestingly rugged-looking diode valves that had been taken from unexploded German bombs. Tests made on them in a centrifuge in the Cambridge Biochemistry Laboratory convinced him it was possible to make valves tough enough to withstand firing from a gun.

Cockroft took the problem to Pye, where B.J. Edwards produced a sample within 6 weeks. Edwards had foreseen the importance of miniature valves in television, and before the war had persuaded C.O. to take surreptitious control of a struggling valve-maker called Hi-Vac. By the end of 1938 Hi-Vac was turning out a series of experimental valves for Pye, and one of these trick valves, as Edwards called them, was the tiny 'acorn' that allowed Pye to produce its first prototype proximity fuze so quickly. In May 1940 Edwards wrote to Cockroft's office about an 'idea we have on a magnetic anti-aircraft shell ... [which would] explode when it is within a certain distance of the aeroplane at which it is aimed'. Within days Forman was in London and impressing Cockroft with his account of Pye's work both on magnetic and radio-operated proximity fuzes. Two weeks later Cockroft took Professor Blackett to Cambridge, where they agreed that Pye should pursue the radio-operated version as the more promising. Pye, Cockroft noted, 'would not require any development contract for this since they preferred to work independently'. There is no record that C.O. was at the meeting, but it is easy to imagine him standing up to make that speech.

Later that June, Jackson test-fired a Hi-Vac 'acorn' from an AA gun, recovered it and took it back to the laboratory and found that it still worked. The next step was to put a fuze carrying a receiver in a shell, and to trigger it off by a pulse sent from a transmitter on the ground when the shell was near the target plane. A government report later commented that 'the urgency of [Pye's] work can be gained from the fact that this transmitter [with] an effective power of 1,000 kW was constructed by Pye within one month'. The final stage would be to put both transmitter and receiver in the shell nose cone to obtain a true proximity fuze.

A Pye independent of Whitehall and using its own resources to make a super-weapon was C.O.'s idea of how the war should be fought, but the new fuze was too important for this idyll to continue more than a few weeks. Later that summer other scientists and firms were assigned to the project. Pye was now to work under Ministry of Supply contract on the fuze circuit and transmitter, but co-operate on the development of valves with GEC, which had far greater resources for valve development. It would also co-operate with W.A.S. Butement on receiver design. By November GEC had built and test-fired a miniature all-glass pentode valve. Pye tests had already established the resilience of valves made of glass but even Jackson admitted that the new GEC valve was tougher than his own. On Christmas Day 1940, Edwards' team completed their first radio fuzes.

It was the start of a trail of disappointments. Later test firings showed that valves and radio circuits were mangled by the shock of being fired from a gun. Every ounce of radio equipment in the fuze had to withstand a force of more than a third of a ton while remaining effective inside a

Figure 4.9   A collection of proximity fuzes

shell spinning 250 times a second. The unsuccessful shells – from the scientists' point of view the interesting ones – also had to withstand the shock of landing. Often they did not and, in the words of the government report already quoted, ended up looking 'as though they had been passed under a steam roller so that post mortems were difficult'. At this point Donald Jackson, as C.O.'s Chief of Mechanical Designs, was charged

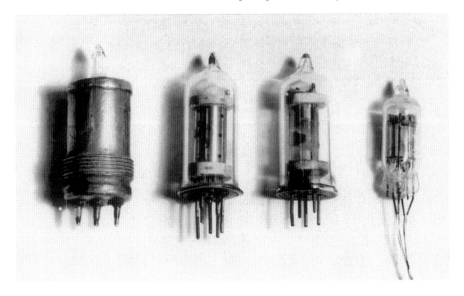

Figure 4.10   Valves designed by Pye for the proximity fuze programme, and tested in recovery shells

with 'making the fuze sufficiently strong to withstand these terrific conditions'.

He was the right man for the job, for if he had a greater passion than planes and fast cars it was explosives. He had first tested Pye's own 'acorn' valves by dropping them onto a lead block from the top of the lift shaft of the University library, enjoying the sight of grumpy old dons being turned away when the library was closed off with security signs. He now carried out tests at Shoeburyness, where officers and NCOs wore blue caps, red jackets and white trousers so that they could be safely seen on the firing range. Jackson spent as much time in Wellington boots as in shoes, for an important part of his job was to walk miles over the sand at low water collecting the remains of shells that were fired when the tide was high. Each week Pye assembled the components needed for more of the dwarf fuzes, and each week Jackson loaded them into his car and set off for Shoeburyness. Safety precautions were strict. The fuze had to remain dead until it reached the target plane, and if the shell failed to go off it was supposed to explode before reaching the ground. Jackson was present when each fuze was screwed into its live shell, which he admitted 'made one a bit careful'.

Pye and GEC could not get the radio fuzes to explode the shells reliably and, hoping to improve the success rate, the Pye mathematician Donald Weighton used an unmanned balloon with a weighted basket to test the strength of signal needed to touch off a fuze. When a batch

*Figure 4.11   Scale model of a German plane made by Dennis Lawson, Pye's Chief
Research Engineer, used to calculate the distance from a plane at which
the proximity fuze would detonate*

of shells was fired on 6 August 1941 one was successfully exploded
by a transmitter pulse. There had been so many failures the scientists
wrote it off as coincidence, but from this time the number of successful
detonations by transmitter increased. At the same time C.O.'s Chief
Research Engineer, Dennis Lawson, made scale models of German
planes and flew them past a tiny model of an AA shell to calculate the
distance from a plane at which a true proximity fuze would detonate.

With his engineers increasingly confident, C.O. invited General Pile
to observe a test shoot of the radio-operated fuze on 11 February 1942.
Three-quarters of the shells were successfully exploded by pulses from
the ground transmitter. 'It was quite obvious', Pile wrote, '[that] we
were already in sight of the complete defeat of any personnel-controlled
aeroplane', and he attributed the success to the work of Pye and EMI,
and in particular to B.J. Edwards' development of the valve small and
strong enough to survive inside a shell. It was now decided that Pye
should collaborate with GEC to make a true proximity fuze which
would use its own tiny transmitter and receiver to explode a shell near
a target plane without the need of a triggering signal from the ground.
Here trouble began. The British proximity fuze, the X3, used a GEC
16 mm diameter valve while the Americans, adding the information

brought by Tizard to their own research, had the advantage of a smaller 10 mm valve. General Pile's confidence was only partly justified: the proximity fuze was about to revolutionise the anti-aircraft gun, but it would be a fuze made in America.

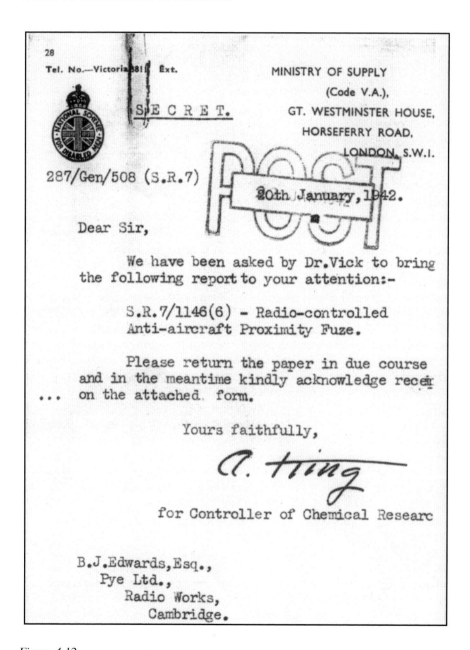

Figure 4.12

When the Tizard Mission began its talks in Washington in the early autumn of 1940 it soon realised that the Americans were working on their own version of a proximity fuze that they called a variable-time, or VT, fuze. This American experience was considerably enriched when Tizard allowed them to copy secret papers covering Pye's early attempts to make a rugged valve, its designs for a radio-operated fuze and, most importantly, Butement's designs for a more advanced fuze based on the Doppler effect. From that time C.O. sent reports on Pye's progress to Merle Tuve, head of the US programme, and also despatched Dennis Lawson to America for several months with details of his own research and a box of Pye's latest radio-operated fuzes. This and other British information helped Tuve focus his research, and the American fuze programme soon dwarfed Britain's. Some 1500 scientists and engineers went to work on the fuze in US laboratories compared to just 50 in Britain, and they were backed up by the facilities of giant companies such as Sylvania and National Carbon for making the valves and batteries that had to be both diminutive and reliable.

John Cockroft, as head of the Army's Air Defence Research and Development Establishment (ADRDE), found it a saddening experience. Britain's quest to make the proximity fuze, he wrote in his post-war paper, *Memories of Radar Research*, 'was a continual struggle to overcome the disabilities due to poor industrial development facilities', and ADRDE's failure with the elusive fuze was its 'greatest disappointment'. In May 1946, a year after the war in Europe had ended, the Ministry of Supply delivered a similarly gloomy verdict: 'very little production as such has been undertaken [in Britain] on VT fuzes'. They had only been made in limited batches for development trials, and the Gramophone Company that was supposed to produce 4000 of the latest model valves each week never turned out more than 1500. By the end of the war America had produced 150 million.

The American breakthrough came just in time. When Germany launched the V1 flying bomb in the summer of 1944, gun-laying radar had improved the effectiveness of anti-aircraft guns, but the V1 was reckoned eight times harder to shoot down than a German bomber. A big delivery of American VT fuzes arrived in Britain shortly after the first flying bombs hit London, and in their first week of use anti-aircraft guns brought down a quarter of the V1s they shot at. By the final week the gunners were destroying four out of every five.

C.O. and his little team were vindicated, but small wonder he felt hard done by. He had told his staff that eventually reward would go where it was due, but this did not happen with the secrets Tizard took to America. As soon as the Washington talks began it was plain that patents were a problem best avoided lest it hold back vital war work. Patent rights of British individuals and companies were temporarily waived, and America developed and manufactured what it liked of the

Tizard treasures, though without invalidating any later British claims. The dispute over the fuze was not settled until 20 years after the end of the war, by which time the technology was antiquated. The enterprise of C.O. and Pye, and of other British companies and government scientists, was used to pay America for putting its resources at Britain's disposal even before America itself entered the war. C.O. never admitted that this was the price for winning the war, nor did he ever concede that the ingenious fuze might not have brought down a single German V1 if Britain had tried to develop it on its own.

An army goes on air

In the late summer of 1939 the Ministry of Supply put out tenders for the British Army's first portable infantry radio. Little money had been spent on Army radio after World War I; the needs of the Air Force and Navy were seen as unarguably greater, and high-ranking soldiers still questioned the usefulness of radio communications. Although the War Office had powerful radio equipment linking it to commands at home and abroad, short-range communication was still often by flags, lamps and even heliograph mirrors. The radios the Army did have were often inadequate. Poor aerial coupling meant they radiated only a fraction of their available power; and all the sets used high power modulation systems that made them bulky and greedy users of their batteries.

The new set, for use between infantry units in the field, had to be light, simple to use and economical. The Army's Signals Experimental Establishment (SEE) came up with the No. 8 set which field trials showed to be heavy and potentially expensive to manufacture but otherwise satisfactory. Pye and other companies were asked to prepare bids to make the set, and 11 days after the outbreak of war Charles Harmer and his engineers sat down to consider the Pye offer. Harmer was now C.O.'s senior voice on radio matters and, since March 1939, a director of the company. He had friends among senior Army officers and got on well enough with civil servants for the post-war Ministry of Supply to offer him a job. At the same time Harmer was in awe of C.O. (and of his sister Pearl, under whom he worked in Invicta) and he always did as he was told, though even he was shocked when C.O. joined the engineers' discussions on the new military radio, as Dennis Fuller recalled.

> [Harmer's group] started rehearsing their response [to the tender] to C.O. and after about half an hour his attention started to wander. He got agitated and . . . suddenly asked how much this radio set was going to weigh and how it was going to be carried. The specification required steel for all of the chassis and casings, and [the set would need] to be carried by two people. This seemed mad to C.O. [and] he told Harmer not to reply to the tender.

According to another account – there was often more than one version of stories about C.O. – Harmer had already identified serious weaknesses in the No. 8 set and did not like the Army's specifications any more than C.O., but he was shocked by C.O.'s imperious rejection of a design produced by soldiers and government scientists whom he considered important people. For C.O., incensed by what he saw as the stupidity of making portable radio sets that were not truly portable, rejection was the only choice, and he sent off a letter to the Ministry of Supply that was both rebuke and challenge.

> We think it is obligatory on us as British manufacturers to state that our opinion on the design and production of this receiver is at variance with the opinion of the other . . . parties. We would consider it dishonest for the sake of obtaining a contract to acquiesce in a policy with which we are not in agreement.

Pye, he went on, could make the No. 8 set, but it could also design and develop a lighter radio that was just as efficient. Production of this set would need less tooling and therefore be quicker and cheaper. He would, he concluded, happily meet with the Ministry of Supply's experts to explain his proposal at greater length, but unfortunately he did not have the petrol for the drive to Woolwich.

This was either cheek or a bravura demonstration of self-assurance. Pye had little experience of military radio, yet other manufacturers equally or better qualified had not objected to the set's specifications, let alone refused to make it. C.O. was not a radio engineer (his staff were never sure how much he understood of technical matters), yet he was ordering an alarmed Charles Harmer to refuse a profitable government contract for a radio designed by the Army's best signals experts.

In the end C.O. chose to use cunning rather than confrontation in his campaign against the No. 8 set. Sam Carn, whose job was to buy the components Pye needed for wartime production, had also heard C.O. say about the No. 8 set 'It's no good. We're not going to make the bloody thing.' Not long after, though, Carn saw his boss make a change of tack. C.O. needed money to develop the infantry radio he thought the Army should have, and so he accepted the Ministry of Supply's contract, worth £25 000, to make the set he swore was no good.

He did not make, and probably never intended to make, a single one. Instead he spent the £25 000 and three or four times as much besides to develop his own set. Pye's alternative, designed by Bob Dalgliesh and Dennis Hughes, weighed 29 lbs compared to the No. 8 set's 50 lbs. Pye could not use aluminium and light alloys, which were reserved for aircraft and other special needs, but C.O. found a supply of tin plate used for canning. Corrugated to add strength, this replaced the heavy steel of the Army's original design. Pye's engineers worked fast, testing two alternative prototypes on the Gog Magog hills outside Cambridge, where police arrested them for making illegal radio transmissions. The sets were ready in less than 2 months and C.O. lobbied the Army to hold trials in which they could be tested against the No. 8 set. C.O. pressed so hard that he forced Brigadier J.A.S. Tillard, then responsible for the production of signals equipment at the Ministry of Supply, to break a fishing holiday in Scotland and return for talks in London. C.O. took Norman Twemlow, one of his pre-war sales staff, to the meeting where Tillard complained it was hardly cricket to make him break his holiday. According to Twemlow, C.O. said 'we aren't playing cricket, and if that's all you are concerned about we may as well give up'. He pushed back his chair and marched out followed by an apprehensive Twemlow. (Tillard forgave C.O., and when the war was over wrote a generous letter thanking him for his 'magnificent contribution', adding 'what we would have done without Pye, I tremble to think'.)

C.O. got his way and in January 1940 field trials were held by the Welsh Guards in northern France. He suspected that the Army chose France, inconvenient to reach and potentially dangerous, in the hope of keeping Pye's engineers away, but Harmer and his team made the uncomfortable journey and were rewarded by the Guardsmen's approval of one of their two designs. Lighter and more compact than the Army's model, it used less battery power and needed only four valves instead of seven; known as the No. 18 set, it became the standard infantry man-pack radio of World War II. Pye put it into production at once, and later it was also made by other radio manufacturers including Ekco, Murphy and Bush. Seventy-six thousand No. 18 sets were produced during the war, and used in almost every theatre including North Africa and South East Asia, as well as on airborne operations. It was poorly suited to some of these tasks and was not, as Pye engineers admitted, a great technical achievement. It was simply the best that could be done quickly under the conditions of the time, but its importance for C.O. and Pye was enormous. A company with little record of working in military radio had challenged, and defeated, both the Army and the rest of the industry.

C.O. later claimed that a senior Army officer lost his job over the No. 18 set. There is no record of this, but it is certainly true that his colleagues worried that his truculence might so anger Whitehall that Pye would never work on another military radio. In fact C.O. had shown that his way of running a company – no committees, no cautious procedures – made it possible to design and develop a set faster than anyone else. This was to be proved again, though on this occasion Pye would make an innovative set that was arguably the most successful radio of the war.

Since the mid-1930s the gifted government scientist W.A.S. Butement had been working on a new concept in tank radio that allowed

Figure 4.13 The Duke of Kent talking to a soldier using a No. 18 set during an exercise

Figure 4.14 The Prime Minister, Winston Churchill, watching two soldiers using a No. 18 set during an exercise

Figure 4.15   Servicing a No. 18 set

Figure 4.16   An officer is handing over a message for transmission by wireless to Battalion HQ

Figure 4.17   A soldier in an advanced party position is transmitting a report of the situation on a No. 18 set

Figure 4.18   Wireless messages being sent and received at a village shop in Northern Ireland, used as HQ during an exercise

Figure 4.19   Dennis Fuller, from Pye's research team, with a handset radio telephone developed by Pye in 1942. The range between two handsets was 500 yards with a battery life of $1\frac{1}{2}$ hours (later extended to 4–6 hours using American midget type valves)

the deployment of armoured formations in deep and rapid thrusts into enemy territory. The first British rehearsals of the tactic that German tanks made famous as *Blitzkrieg* were conducted on Salisbury Plain, with tank sets operating on the same frequency. After the British retreat from Dunkirk in June 1940, Pye was asked to develop a radio for use in armoured vehicles. This time the Ministry of Supply gave Pye written specifications, leaving the firm to make the drawings and work out costings. The motive for asking Pye may not have been straightforward: after the row over the No. 18 set it was merely prudent to offer the opportunity to a company with such talent for trouble-making. There were other designs for tank sets to choose from, though one was clumsy and expensive, another a poor performer, and neither was suitable for mass-production.

The offer put C.O. in a difficult position, as perhaps some in Whitehall hoped. Dalgliesh and Hughes, designers of the No. 18 set, were fully occupied, as were other experienced engineers. C.O. took soundings and called in the 23-year-old Bill Pannell, who had joined the company as an apprentice. Pannell had helped out on the No. 18 set but was now working on transformers. He had never designed a radio of any kind, let alone a set for modern tank warfare. He described the unexpected meeting with his employer.

"Well Pannell," [C.O.] said. "You're our best man for designing radio sets for the Army. This new contract could be very important for the company. We're pinning all of our hopes on you." I suppose I was always a quiet sort of chap, and I didn't say much – didn't really know what to say. I did tell him I had never designed a set from scratch before but would give it my best. "You're a good man Pannell, one of our best – we're depending on you." Not many of us got to see the boss like that – I was very pleased, proud.

No one had ever spoken to the unassuming Pannell like this, and Dennis Fuller recognised it as C.O.'s way of 'inspiring and encouraging young, inexperienced people to achievements that they would never have dreamed of'. Pannell did not work alone on the set, and had more experienced colleagues to turn to for advice, but few other companies would have given him such responsibility. C.O.'s procurements chief Sam Carn liked to say Pye could 'beat hands down [firms] steeped in the government attitude' that were horrified by the way Pye dodged official procedures in order to develop equipment fast. The radio Pannell produced, the No. 19 set, consisted of a main transmitter and receiver, the A set, for communication between a tank squadron and its regiment and a B set for close-range communication between tanks within a squadron. There was also an intercom that allowed the commander to communicate with his crew (previously signalling between tanks was by flags; between a tank crew by shouting). The work was so urgent that Pannell made and tested the B set by September 1940. The specifications for the A set called for the use of two frequency bands but when this delayed development Pannell was allowed to make a preliminary model using a single frequency. Final trials were held in

November 1940 and Pye was given orders for 3000 Mark I No. 19 sets, while Pannell was instructed to work urgently on a Mark II that gave the A set its full frequency range.

Testing was a key part of Pannell's work. On visits to Aldershot and Catterick he noticed that soldiers seldom handled equipment in the way its designers intended, and that tank crews often used their radios as steps to stand on. 'That was what made me put . . . very strong carrying handles on either end of the set for the first time. They were partly carrying handles, [but] they were partly handles that soldiers could stand on safely without bending or breaking radio knobs.' Pye aimed for practicability, which explained why C.O. was such a demon for testing. According to Fuller '[he] didn't trust the senior Army people at all – thought they were all asses', and he had trenches and fox holes dug in the playing field beside the factory so radios could be tested in realistic conditions. Sets were dropped on tarmac and in puddles, and operated from behind, and on top of, buildings. Whenever a storm blew up there was a rush to take sets outside to see how bad weather affected reception and whether they got waterlogged. When making a radio that had to be transportable on a mule or by vehicle the engineers put one on a donkey and had it run round Midsummer Common to see if it fell off – 'lots of fun', Pannell thought. Little of this work was logged or methodically monitored. If a detail or part was wrong, or showed a weakness, it was corrected: 'unscientific', according to Dennis Fuller, 'but generally effective'. Fuller had already tested his own prototype hand-held radio for platoon commanders – it was one of the treasures Tizard took to America – and was convinced that Pye 'taught the Army more about testing [radios] and how to specify new sets than they ever knew'.

C.O.'s suspicion of the Army was confirmed by trials that pitted Pye's tank set against three rival prototypes designed by the Signals Experimental Establishment. Accounts of what happened are sparse, but the Chiefs of Staff declared one of the SEE sets the winner, and then asked C.O. to make it. He refused to touch what he called a bad set; further trials were held in which two of the War Office sets proved unserviceable, while Pannell's radio performed well. Whether the results of the first test were muddled by mistake or on purpose there is no way of knowing, but letters from senior officers in the Tank Corps thanking C.O. for the No. 19 set allude cryptically to 'an inner history of your battle for us in connection with [this set]'. And several years after the war Sir Geoffrey Burton, a senior official in the wartime Ministry of Supply, reminisced with C.O. about the 'events leading to that phoney test . . . Harmer (in overall charge of the No. 19 set as of all Pye's radio projects) really ought to write a thriller about it. It would be a best-seller'. It is indisputable, though, that some in Whitehall were furious about the No. 19 set's victory and continued arguing even after the war

Figure 4.20   Two members of the ATS (Auxiliary Territorial Service) assist in fitting a No. 19 set into a tank

that Pye won the contract for 'political reasons' rather than technical excellence.

The authority on military signals, Brigadier J.B. Hickman, called the No. 19 set a 'landmark in the history of military communications' because it converted soldiers to the full exploitation of radio. The No. 19 set's success was a triumph for C.O. and Pye. Burton, in a letter written shortly after the war, told C.O.,

> if it had not been for all the enthusiasm you showed in the early days for developing a set that could be produced quickly and would work, I do not know where we should have been so far as the tank set was concerned.

No other army of World War II had such a versatile radio and by the end of the war 2 million No. 19 sets had been made in Canada and the United States as well as by several companies in Britain. Pye alone produced 115 000, including 3500 for the Red Army which, the Ministry of Supply told C.O., had no complaints about its operation or maintenance. A letter to C.O. from a serving officer in the Royal Armoured Corps suggests the soldiers' attitude: 'your firm has made

Figure 4.21   Giving instructions on using a No. 19 set

one of the greatest contributions, if not the greatest, to the equipment of
the [Corps]. We are still trying to get better tanks and better weapons,
but with our wireless set we are satisfied.'

In spite of this success, but also because of it, C.O. was making more
enemies. In radar he had in Robert Renwick a guardian angel subtle
enough to appreciate his useful qualities and ignore the rest. No such
heaven-sent creature watched over C.O.'s dealings with the Army and
the Ministry of Supply, and it was not surprising that some officials felt
that Pye should suffer for its affronts to Whitehall's code of conduct. In
The Challenge of War, Guy Hartcup, a wartime member of the Ministry
of Supply, argues that the story of the No. 19 set illustrated the clash
between War Office values and those of commercial firms in general
and Pye in particular.

There were two extremes ... the highly cautious War Office policy
leading to expensive but reliable equipment in small numbers because of
the price and effort involved; [and] the hasty, rather slapdash approach
of the commercial radio company, as in the case of the 19 Mark I,
leading through a series of models to a reasonable equipment.

This 'rather slapdash approach' was successful in the development of radar, where the need for instant production and constant adaptation to an improving new technology soon made equipment obsolete. C.O.'s quick-footed leadership admirably fitted Pye for that sort of work. The slower-moving Army did not easily take to the idea of designing equipment for almost immediate, large-scale production and removing the weaknesses that were the inevitable result in later models. This was what happened with the No. 19 set, whose Mark I was found to be dangerously flawed in North Africa in 1941 when the radios of British tanks failed in the first battle of El Alamein. The set's condensers had been designed to work at maximum temperatures of 70°C, and when operated in the desert at above 100°C they packed up. More suitable condensers were flown in from America and the fault was corrected in later models. But it is hard to see how any other method of development would have allowed such rapid production of a revolutionary radio.

After his experience with the No. 19 set, C.O. was on the look-out for any hint that Pye was being treated unfairly, and he thought he detected something suspicious in the events surrounding the creation of another multipurpose radio, the No. 22 set that Pannell tested on the back of a donkey. In February 1942 C.O. heard from an acquaintance in the Ministry of Supply that an official had said Pye was so overloaded with work it could not fulfil its contracts with the Ministry of Aircraft Production. His source for what he insisted was a grossly damaging slander was William Dwyer, an Irishman who had volunteered to join Britain's war effort and worked in the section of the Ministry that procured equipment for tanks. He was also an old friend and business colleague of C.O., who some ten years earlier had bailed out the Dwyer family's textile business in County Cork.

On the strength of Dwyer's information C.O. wrote an indignant letter to the Ministry of Supply denying that Pye was swamped by contracts. If his company could not work to full capacity it was because it needed more labour, for the government had made it run a three-shift system without understanding that there were not enough workers to operate it efficiently. Any delay in Pye's deliveries was caused by shortage of components; 'overloading' had nothing to do with it. The Ministry's Deputy Director General, Signals Equipment, Major General D.A. Butler, commented on this letter in a minute: 'Mr Stanley means his firm is only working at 50 per cent of ... capacity; he is always saying this, though judging by the lateness of some of his deliveries, it would appear that the firm is overloaded.' Pye certainly was behind in its work for MAP, Butler went on, but was on time with production of the No. 19 set, and he admitted that late deliveries by subcontractors contributed to Pye's difficulties.

C.O. knew nothing of Butler's comment and asked him if it were true that someone at MAP had said Pye was not keeping its 'production

Figure 4.22   Using a No. 22 set fitted into a Humber truck

Figure 4.23   Servicing a No. 22 set

promise'. If so, he wanted to know who, and also to see all relevant papers on the matter. An unimpressed Butler answered that the supposed remark was 'so vague ... it will be extremely difficult to trace'. Any chance that C.O. might be fobbed off vanished when he remembered a visit by a colonel from the Ministry of Supply while Pye was in the early stages of its work on the No. 22 set. The colonel told him the Ministry felt it was 'getting too much into the hands' of Pye and wanted C.O. to collaborate with Ekco on designing the set before letting Ekco take over its production. C.O. resisted this, but then learned that while Philips had been asked to produce the No. 22 set in large quantities, Pye was to be given an order for only 100 sets a week. Interpreting this as part of a plot to down Pye, he asked Butler to arrange a meeting with his Minister.

C.O.'s tactics veered from furious demands to declarations of patriotic self-denial. He had not, he told Butler, 'taken exception to the fact that someone else (ie Philips) got a contract and we did not. If it is in the interests of the country, I feel that I have demonstrated to your Ministry on many occasions that I have foregone any advantage to my own company.' But he was sure the remark of the man from MAP had caused Pye to get a smaller contract, and 'I want to find out who that executive was because I consider that if the action of one Ministry is guided by a report from another ... it should be stopped.' Butler again tried to fob him off ('nothing useful could be gained' by trying to track down the troublesome remark), but C.O. would have none of it. If Butler could not find the author of the poisoned words would he allow C.O. to ask William Dwyer? Butler called in Geoffrey Burton, to whom C.O. was better disposed, and he too advised him to drop the matter. The words, if uttered, were at worst a half-truth, Burton wrote soothingly, but 'malicious', and the Ministry did not allow itself to be influenced by 'chance remarks'.

This brought C.O. to a new pitch of anger. The objectionable words had appeared in a 'high level' minute, he told Burton. The Ministry of Supply had informed Philips, who passed it on to C.O., that Pye was not being given work because of overloading. 'Important public principles are involved. This matter should be dealt with openly and without delay' and he wanted to register 'a vehement protest at the treatment I have received'. Early that summer Dwyer lost his job. The Irishman suspected he was being punished for his part in the affair and told C.O. about a discussion on the No. 22 set in January 1942 in which Brigadier Tillard (who later thanked C.O. so warmly for the No. 18 set) twice said Pye was 'overloaded with contracts'. Dwyer's letter, which seemed contrived to show C.O. as the injured party, included a theatrical apology for having believed the criticism he heard of Pye and 'for the way I treated your staff when I first took over [my post in the Ministry]. I was completely misled regarding the quality of [the

No. 22 set] and the grand work that your firm have done.' C.O. passed this 'astonishing' revelation to General Butler, saying it justified him having questions asked in the House of Commons.

He was discovering how useful Members of Parliament could be, and if he lost interest in the 'slanderer' in the Ministry of Aircraft Production it was perhaps because he had found bigger prey. The Ministry of Supply, and to a lesser extent the Air Ministry, provoked him as a rabbit warren excites a terrier, and he found their cumbersome procedures and slow-moving personnel an irresistible hunting ground. In March 1942, while C.O. was snapping at General Butler's heels, the deputy prime minister Clement Attlee told Winston Churchill that he had been approached by MPs worried about the state of radio and radar equipment in the Armed Forces. Attlee knew the names of the MPs' informants and considered they 'inspired confidence', not least because their reports came from 'quite different sources which bore out the same conclusion'. Churchill quickly set up an inquiry under Lord Justice Herbert du Parcq to study three matters: co-ordination between the services on radar; the adequate use of technology in communications between Army units, and between planes and tanks; and 'whether the situation with regard to the production and supply of wireless equipment to the three services is satisfactory' (in his letter to Churchill, Attlee had alluded to damaging 'quarrels and jealousies').

C.O. later talked as though he single-handedly brought about the du Parcq inquiry and, though the Attlee letter makes plain this was not so, there is no doubt he helped prepare the ground for it. Du Parcq heard 62 witnesses, among them officers from all three services, the cream of government scientists starting with Lord Cherwell, Tizard and Watson-Watt, four MPs, and just one representative of Britain's electronics industry – Charles Harmer of Pye. His participation suggests that C.O. was most interested in the inquiry's third subject, the production and supply of radio equipment, and this was where du Parcq's judgement was harshest. 'Nobody', he concluded, '[had] been prepared to answer' that this was satisfactory. Lack of components was the chief problem, he said, quoting Cockcroft to the effect that this was 'the real bottleneck'. Du Parcq recommended a high-level Radio Board to supervise radio and radar production (part of its brief was to pursue greater standardisation of equipment and components), and the following year C.O., apparently satisfied, told Pye directors that there had been 'a great number of changes in high places'.

His battles in these 'high places' led to some improvement in contacts between Pye designers and Army signals experts who, so Pannell thought, 'became much better at knowing what they wanted'. But just as the No. 22 set was going into production in the summer of 1942 a new campaign against Pye and C.O. Stanley was being prepared inside the Ministry of Supply. Ian Gray, the Ministry's new Deputy

Contractors Name and Factory Location / EQUIPMENT	Contract No.	Service Dept	Page No.	Month	Quantity on Contract	Deliveries To Date	Deliveries June	Forecast June	Forecast July	Symbol "2" if Deliveries forecast to be due in (Aug. Sept. Oct. Nov. Dec. Jan.)
POWERED MOUNTINGS - ASSOC. PARNALL AIRCRAFT TOLWORTH (CONT'D)										
SCANNING U 83	SB 50420	31	25	06	2250	1247		225	225	22221
SCANNING U 92	WT 14401	31	17	06	72	72	72			11111
SCANNING U 106	WT 16160	31	21	06	100				5	22222
STABILISNG PLAT 26A	SB 56503	34	19	06	2975	1327	279			11111
STABILISNG PLAT 26	WT 14879	31	19	06	2000					11111
POWERS & CO CROYDON										
LEADS PLUGS DOUBLE	SB 51577	33	19	06	64002	63942		60		11111
LEADS PLUGS DOUBLE	SB 51577	33	19	06	99999	99999				11111
LEADS PLUGS DOUBLE	SB 51577	33	19	06	99999	99999				11111
PLUG BOARD	SB 51577	33	19	06	17600	13650	1000	1200	1200	22211
TYPE X MK VIA	SB 26109	33	17	06	1900	208	64	50	80	22222
PRECISION ENG WELLINGTN ST LEICS										
GEAR BOX 36	WT 11125	34	7	06	100	67	27	60	33	11111
FREETHARE MERTON S W 19										
INSERT BODIES	WT 11298	33	[illegible]	06	6300	3735	1800	1200	1200	22111
INSERT BODIES	WT 12629	33	18	06	8700					12222
PLUNGER BODIES	WT 11298	33	18	06	23000	14720	960	2000	1000	22221
PLUNGER BODIES	WT 12629	33	[illegible]	06	6000					11122
PRIDE F H CLAPHAM S W 4										
CONTROL U ETC	C 61015	10		6	100				15	222222
CONTROL U ETC	C 67527	10		6	300					111111
RECTIFIER U S E S	CI 9931	10	1	06	1250	948	125	140	100	222111
PUTTICK T B HASLEMERE SURREY										
CALIBRATOR U	C 71344	10	1	06	230	214		16	16	111111
CATHODE FOLLOWER U	CI 7397	10	1	06	500	326	33	100	50	111111
SWITCH U 14	C 90096	10	2	06	1000	925	425	500	75	111111
PYE RADIO CAMBRIDGE										
REC P11 ETC	CI 1867	10	2	06	40	30	5	12	10	111111
REC P 24	C 72029	10	1	06	540	349	43	50	50	222111
WAVEMETER G 73	C 60383	10	1	06	500	172	68	50	50	211111
APP L 8 9 LOUD HAIL	P 16399	22	12	06	526	6			20	22221
F A NO 2 MK I	SB 6040-0	23	3	06	43	2		2	2	22222
P C R I	P 11442	22	9	06	5000	4979				11111
P C R 2 AND 3	P 15029	22	9	06	12000	1560	760	880	800	22221
W S 10	P 1895	22	1	06	400	304	13	16	12	22222
W S 18T	P 14311	22	1	06	2000+	99	99	[illegible]	400	22211 + Cancelled 13.8.4
W S 62	P 15352	22	4	06	8800					11222
W S 62 MK I	P 13273	22	4	06	5060	49[illegible]4	1260			11111

POW - PYE - SHEET 42

Figure 4.24 Ministry of Supply schedule of contracts with manufacturers (including Pye Radio) of military radio and radar supplies

Figure 4.24   Continued

Director of Signals Research, took up his job convinced, like some of his colleagues, that the procurement of radios for the Army was being mishandled, and he carried out a lengthy study to prove it. His findings were that while the Army's radar sets, mostly for gun control and few of them made by Pye, were 'well engineered', its radios were 'shoddy stuff'. They lacked 'sound engineering design' and were 'far too heavy for human porterage' (one wonders if he knew the history of the No. 18 set).

The trouble, Gray decided, was that radio design 'had been wrested from the Signals Research and Development Establishment and had passed largely into the hand of Messrs. Pye and to a lesser degree Messrs. Murphy'. He cited as evidence for what he called Pye shoddiness the condenser failure in the No. 19 set in North Africa and the fact that the original No. 18 and No. 19 sets underwent several model changes. After the war Sir Robert Watson-Watt pointed out that Britain had been able to produce its early radar sets because firms had been ready to 'sink [their] pride' and cut back on quality in order to satisfy the needs of crash programmes. Gray thought such methods belonged to an 'unsavoury past', and set out on what he called an 'openly declared policy of controlling the final [Army radio] designs from the Ministry'.

At the beginning of 1943 Gray told Harmer and other industry engineers that the Ministry planned a new range of military radios whose specifications would be set by his design team and not be open for discussion. This triggered an angry memo to the Radio Board from Pye, written under Harmer's name but with C.O.'s fingerprints all over it, accusing the Services of giving orders to the industry's communication engineers rather than consulting with them. In May 1944, C.O. told the Minister of Supply that the new procurement policy had produced no equipment of value and that his Ministry had ignored the radio industry's warning about shortage of components and the advice of its most knowledgeable engineers. The situation was so bad, he concluded with a flourish of patriotic pathos, that at Cambridge 'engineers with experience of military requirements – in order to help the Army – have had to continue their development work under cover'.

When the Ministry of Supply's designers launched a counter-attack on Pye's record and demanded a public enquiry to clear their name, someone in Whitehall seems to have decided that the row was getting out of hand. A high-level memorandum was circulated complaining that too much time had already been taken 'dealing with Mr Stanley's allegations', and pointing out the risk of setting a precedent by holding an enquiry 'merely because an industrialist criticises a member of the Ministry'. A later demand for an enquiry from C.O. was dismissed as 'an impertinence'.

C.O. was particularly upset by what he saw as the Ministry's mishandling of the Pye 62 set. This was a variant of the No. 22 set, with the same power but meeting new Army requirements for a radio that was waterproof, weighed only 30 lbs, and could survive a parachute drop. Pye was given 6 months to design and produce the first 50 models. The waterproofing, a first for British military radio, was achieved by Ladislav Lax, a Jewish refugee from German-occupied Prague where he had been a professor of mathematics. Rabelaisian and unconventional, Lax was the sort of scientist who relished working with C.O. He built a tunnel of water trays heated by light bulbs where he tested his radios for resistance to humidity, later taking them to the Elizabeth Bridge near the Pye factory where he threw them into the Cam to see if they kept out the water.

C.O. gave his version of the Pye 62's development to the Minister of Supply, Sir Andrew Duncan, at a meeting in January 1945. The Ministry had rejected the Pye 62 set 6 months after Pye started work on it, and it would never have reached production if General Butler and Sir George Lee, both senior figures in the Ministry, had not visited Cambridge and given it their personal approval. This radio was used in jungle warfare in Burma, in the Normandy landings and the Rhine crossings. C.O.'s team later adapted both the No. 22 and Pye 62 to become Britain's first 'scrambled' (i.e. secure) combat radios, and they were used for command purposes by Field Marshal Montgomery and his staff during and after D-Day. No. 22 and Pye 62 sets were dropped with the British 1st Airborne Division in the mismanaged assault on the Rhine bridges at Arnhem. When the 1st Airborne was virtually destroyed some blamed the poor operation of the radios. Today blame is more usually placed on Montgomery and hasty planning. Pye had little more than a week to prepare and adapt the sets, and was never briefed on how and where they were to be used. The Arnhem battlefield extended over a large area and was heavily wooded, conditions the planners had not foreseen and which Pye had not been told to take into account. It was, though, a bruising moment for C.O. and his engineers. 'Terrible it was', Pannell said, '[those] complaints about whose fault it was that the radio sets would do only what they'd been made to do'.

Perhaps to hit back at C.O. for his trouble-making, the Ministry, or the anti-C.O. lobby in it, ordered a check on Pye's production performance, but the resulting report only provided more ammunition for Pye. It was currently making seven different sets and two other major pieces of radio equipment for the Army; production was on or over target except in two cases where, according to the report's author, the delays were beyond Pye's control. He added,

> I regard this firm as one of our best – they give little trouble and almost invariably keep their promises. They are undoubtedly the only firm in the country who can combine large scale production and the flexibility to cope with rather large type diversity.

Two weeks later Ian Gray, now Director, Signals and Radar Development, sent off his own answer to C.O.'s criticisms. Part Jeremiad, part anathema against the man who so tormented him, it went back over the 2 years of what he saw as his righteous war against Pye. There had been moments, Gray wrote, when he 'fervently wished' he had never taken on his new post and 'had stuck to my first job in the Ministry – Bomb Disposal. It was at least good clean fun'. In the end, though, he had brought to heel all the barons of the radio industry with the exception of Pye, where 'I had little success'. He had therefore punished the firm for its 'unco-operative attitude' by giving it none of the Ministry's new radio work. Gray implied that C.O. was angry because he could no longer ignore the Ministry and deal directly with the Army as in the past. 'The issue at stake', he ended,

> is not the one Mr Stanley would so naively have one assume, rather it is "shall the Army have wireless sets designed and produced by Messrs. Pye or shall the Ministry of Supply assume its rightful control of design?"

When the Minister received C.O., accompanied by Harmer, in January 1945 he seems to have tried to soothe him, perhaps recognising he could be troublesome and useful in equal proportions. Harmer did his bit too, politely denying suggestions that Pye wanted to cut out the Ministry of Supply and deal direct with the Army. At the end of the meeting the Minister suggested that the problem was 'lack of good feeling between officials of the Ministry and Messrs. Pye'; had he known this earlier, 'the matter could have been put right long ago'.

That, perhaps, was as close to the truth as it was prudent to go: there was no point addressing the fundamental problem that C.O. wanted to fight the war on his own terms and in his own way. There are no traces of later disputes of such intensity, perhaps because new appointments in the Ministry brought forward people with whom Pye worked well. Officials were able to co-operate with C.O. and Pye, but only if they had imagination, patience and a strong sense of humour. Anyone lacking these risked becoming an enemy, and as Mr Gray discovered, there was nothing C.O. enjoyed more than tormenting enemies. Of all the weapons that came out of Cambridge during the war none needed more careful handling than C.O. Stanley, and for this a price would have to be paid.

Sources

Seeing the enemy

COS files: 1/8/2 (including Stanley's notes on EF50 and removing stocks from Holland), 2/3, 6/1.

Public Records Office – documents in the AVIA, HTT and T series (see Source materials section, pp. 337–44).

Interviews: James Bennett, Brian Callick, C.J. Carter, Professor John Coales, Mike Cosgrove, Richard Ellis, Jo Fletcher, Dennis Fuller, Donald Jackson, Fred Keys, Bill Pannell, Geoff Peel, Peter Threlfall, Gordon Maclagan (view from MAP), Willie Wakefield.

Michael Bell interview: Les Germany.

A–Z files: Professor John Coales, Donald Jackson, Sir Edward Fennessy.

Books: Bell, *ibid.*; Bowen, *Radar Days*; Cockroft, *General Account of Army Radar, Memories of Radar Research*; Oatley, *My Work in Radar*; Sayer, *Army Radar*; Watson-Watt, *The Evolution of Radiolocation*.

The deadly fuze

COS files: 1/8/2, 2/3 (General Pile's 1941 letter), 2/6 (Stanley's history of Pye and the proximity fuze), 6/1.

Public Records Office: documents in the AVIA series (see Note on Sources).

Interviews: W.G. Allen, Richard Ellis, Dennis Fuller, Gordon Maclagan, David Schoenberg.

A–Z files: Guy Hartcup, Donald Jackson.

Books: Bell, *ibid.*; Cockcroft, *ibid.*; Hartcup, *The Effect of Science on World War Two*; Jones, *Reflections on Intelligence*; Pile, *Ack-Ack*; Watson-Watt, *ibid.*; Zimmerman, *Top Secret Exchange*.

An army goes on air

COS files: 1/8/2, 2/3, 2/6.

Simoco boxes: Box 1 (agenda of Pye board meetings 1940–1943, including Stanley's 1943 report on Pye at war).

Interviews: Professor John Coales, Tony Cowley, Don Delanoy, Richard Eden, Dennis Fuller, Fred Keys, Norman Leeks, Gordon Maclagan, Bill Pannell, David Smith, Peter Threlfall.

Michael Bell interviews: Sam Carn, Charles Harmer.

A–Z files: Walter Farrar, Louis Meulstee, Bill Pannell (paper on Pye wartime military radio), Lt Col P.A. Soward.

Books: Bell, *ibid.*; Hartcup, *The Challenge of War*.

Chapter 5

The fighting factory

Two days after Britain declared war on Germany the board of Pye voted to give C.O. a 'free hand' in all 'government work'. Before a year had passed Charles Harmer suggested that the directors should make it plain that this 'free hand' was 'intended to cover the operations of the company's business as a whole . . . in the present abnormal circumstances of national emergency'. There were no objections. Just as war strengthens the authority of a state, Pye's plunge into the unpredictable business of equipping the fighting services served to justify C.O.'s domination of his company.

Wartime conditions meant that directors who did not live in Cambridge (Polson, Ellis and L.G. Hawkins) grew out of touch with developments at Pye. Milward Ellis, complaining of the difficulty of travelling to Cambridge, proposed that full board meetings be held only to discuss the most important matters. Otherwise directors could keep in touch by letter, leaving it to those who lived in Cambridge (C.O., Harmer and Pearl) to 'resolve' matters. Nevertheless Ellis still sometimes raised inconvenient matters. In the middle of C.O.'s battle with the War Office and Ministry of Supply over military radio in 1943 he asked about 'the relationship between the company and government departments'. C.O. said it 'was good with the people who mattered'. Ellis made a last gesture of independence when in 1943 C.O. proposed appointing his brother Eddie to the board. Ellis asked if this was not 'too much family representation'. C.O. dismissed the thought, and said his brother would have to stand for re-election at the next annual general meeting. Unanimously appointed to the board, Eddie remained there for the next 20 years. The Cambridge directors were now three Stanleys, and Charles Harmer, who never contradicted Pearl, let alone C.O.

Eddie replaced another dependable boardroom ally who had gone off to the war. Alan Bradshaw was a former Royal Navy commander who joined Pye as company secretary in the beginning of 1939 and then

Figure 5.1   C. O. Stanley with Commander Bradshaw on left of picture (shortly to be appointed Director of Administration at the Government code-breaking centre at Bletchley Park) just before World War II

became a director. Bradshaw knew little about accounting or company law, but his record as an administrative officer in the Navy was excellent, and when war broke out he was mobilised to administer of the secret code-breaking establishment at Bletchley Park. At first glance he had little in common with C.O. apart from golf. Bradshaw was a linguist, an expert on old silver, and a keen racing man; there was little hint of the entrepreneur about him, and some at Pye thought him rather lazy. But he, too, was born and educated in Ireland; he was also upright and dependable, and C.O. seemed to feel he could trust, and relax with, him. C.O. also established a dependable, though in this instance quite unequal, relationship with Fred Keys, Bradshaw's successor as company secretary. Born into a working-class family in Sunderland, Keys started work at 14 as an office boy before studying accountancy. He was short and inclined to be pompous but (as Keys said) C.O. 'felt comfortable' with him. Keys became guardian of many Pye secrets.

When war began, C.O. sent his mother to live at Lisselan, the house he had bought in County Cork shortly after floating Pye as a public company. His youngest sister Rue went with her, while her husband Bill Tayler looked after the surrounding estate. C.O. travelled to Ireland almost every month throughout the war to see his mother, another sign of his uncontested leadership of the family that he was now establishing throughout Pye. C.O. and Velma rented a house in Cambridge to be near the plant, while Pearl moved down from the London offices of Invicta. C.O. still made frequent trips to the capital. 'People don't work in London', he liked to say, 'they manoeuvre', and he had plenty of manoeuvring to do. On London days he dressed like any smart City gent, but in Cambridge he could almost look like a tramp, wearing a battered hat, old clothes and in winter a ragged racoon skin coat that reached to his ankles. He drove an Austin 10 to save petrol and sometimes tried a bicycle, though he never quite recovered the boyhood knack. He had a habit of turning up at the factory as people came to work and greeting them by name, and often walked the shop floor during night shifts, stopping to talk to people as he went. When word spread in 1939 that the company owned something of great national importance in its stock of EF50 valves the atmosphere in the factory was electric. Audrey Darkin, who ran personnel, felt 'a constant sense of tension, occasionally danger, always a sense of urgency and excitement'. It might have been a description of the boss himself.

In spite of C.O.'s pre-eminence he still needed other people, though he would probably not have recognised his dependence on his brother Eddie, for it was of a subtle kind. Eddie was unassuming (he sometimes made tea for those working the wartime night shifts) and thoroughly decent and people loved him for it. C.O. knew how to stimulate those who worked for him, but he could also be a bully and there were times when he was feared. Eddie was unfailingly gentle; the word most often used to describe him was 'gentleman'. He had 'beautiful manners', Dennis Fuller thought, and 'deserved credit for charming people C.O. had upset. He always had time for [anyone] with a problem, and probably did more to support C.O. than we ever knew.'

No one was left in doubt about what Velma did. The temperament that attracted C.O. to her may have been sharpened by her inability to have children. Before they married she told him she wanted six. 'No promises', C.O. said, and made a grudging offer of one or two. Velma consulted doctors, but fertility treatment continuing into the early 1940s brought no result. War gave an outlet for her energy. She never intruded on C.O.'s running of the factory, but made it her job to see that its employees were properly fed in spite of rationing, bringing to the work the same care for detail she showed in managing C.O.'s domestic life. She ate at the canteen to check on its cooking, and arranged for those who wanted it a cup of Horlicks in the morning and at night. She may have been behind C.O.'s decision to plough up half the sports ground and put it down to potatoes, his Irish pigs and her hens. This was unpopular, for the sports ground and the club attached to it were the heart of the factory's social life, though C.O. kept up the tradition of the

Figure 5.2   Pye sports' day and the tug of war team with their trophy, c.1935. Note the 'rising sun' motif

Figure 5.3   Pye employees on an outing, c.1935

annual sports day (Velma presented the prizes) just as he still arranged works' summer outings to the seaside.

The Pye Home Guard was set up on the first day of war and grew to a strength of 300. The unploughed part of the sports ground was used for bayonet practice, with sandbags hung from goal posts to represent human bodies. It was also the scene of a scandal involving, perhaps inevitably, Donald Jackson, himself a lieutenant in the Home Guard. Returning to Cambridge in good spirits after a day testing prototype proximity fuzes at Shoeburyness he thought it would be fun to paint a swastika on one of C.O.'s sports ground pigs. There was uproar when someone spotted the offensive animal. Police arrived to interrogate staff who might be suspected of sympathising with Germany. An inspector from the RSPCA came to examine the pig. The culprit owned up and C.O. gave him a violent dressing down, allowing Jackson to boast that 'even those who just heard it turned pale'.

The night-time fire-watching rota included C.O., Eddie and Velma, though the large amount of wood in the factory buildings made it likely that just a few incendiary bombs would set off an unstoppable blaze. When German planes interrupted a board meeting in July 1940 no one took to the shelters. The meeting adjourned, the minutes read, 'on receipt of an Air Raid warning as some of the directors and the acting secretary (Fred Keys) had important ARP duties to perform'. C.O. claimed that the Germans planned an early attack against Pye, but

Figure 5.4   The Pye Home Guard

it never came and the city of Cambridge only suffered two raids and limited damage.

The raids were enough to start a duel between Velma and the ARP which almost rivalled her husband's battles with officialdom. Velma drove one of the city's ambulances, and after the second, and worst, of the raids in February 1941 she wrote a five-page report criticising the performance of the emergency services, and stating her 'emphatic opinion' that the Town Hall should pass control of the services to those who worked in them. She also wrote to the alderman in charge of the ARP describing the 'scandalous lack of organisation' during the February raid and the incompetence of the first aid men, one of whom she called 'charming' but 'child-like', another 'unemployable'. She thought Cambridge should learn from the good practice she had observed during a day watching operations at the ambulance depot in London's Berkeley Square.

The aldermen of Cambridge did not like being lectured, and dismissed her. Velma called in C.O.'s London solicitor, C.O. lobbied local MPs to ask questions in the House of Commons, and a public committee was set up to argue that Mrs Stanley's dismissal was a great scandal. In the end no lesser person than Lord Cranborne, deputy regional commissioner for the emergency services, had to hold an inquiry. He took evidence from all the interested parties including Velma, to whom he showed little sympathy. Implicit in the accusation against her, he said in his judgement, 'was the fact that Mrs Stanley was temperamentally unsuited to her job, as not only had she to get on with her inferiors but also with her superior officer'. She had forgotten, he concluded, the ARP's tradition 'that all obeyed orders without question – even if those orders were wrong'. Lord Cranborne did not live in the same world as the Stanleys.

It was another act of insubordination, this time on C.O.'s part, that confirmed the family feeling of the factory at war. Threat of German raids led the government to propose that Pye and other firms engaged on vital war work should move to 'shadow factories' both as a precaution against bombing and to boost production. 'Strong pressure was put upon us', C.O. said when the war was over, 'to build at [government] expense quickly and regardless of cost a giant shadow factory ... somewhere in an industrial area.' Donald Jackson had a less guarded version of this confrontation:

> Funny little man comes down one day [from the Ministry] and tells C.O. that national interests require Pye to move its manufacturing plant to a more appropriate site. 'Where?' asks C.O., "Swansea" this little man tells him "and we will be making arrangements very soon." B.J. Edwards said he'd never seen C.O. in such a state. The arguments went on for weeks – C.O. didn't want to move to Swansea for anyone.

Fred Keys, whom C.O. had quickly taken into his confidence, saw he was horrified by the prospect of 'being lumbered with a factory in Swansea at the end of the war'. He had based his marketing on Pye's association with Cambridge, 'centre of scientific research and excellence'. What was there even for him to sell in 'Pye of Swansea'? And how, if he lived in Wales, could he travel so often to London for the 'manoeuvring' on which he set such store?

Whitehall was adamant and it seemed at least part of Pye would have to move until C.O. had what he called his flash of inspiration. Moving would be a disaster, he told officials, because it 'ignored [the] native reticence [of Cambridgeshire workers] and [their] reluctance to be dragooned'. Instead he promised to increase Pye's production by taking on local sub-contractors and by exploiting the East Anglian hinterland where there were 'a large number of people with skilled knowledge and ability'. Supervised by Pye, but 'working in their own surroundings', these villagers and small townspeople would within 18 months produce as much as any shadow factory.

C.O. convinced Whitehall, and he kept his word. Within a year Pye had 14 000 full-time outworkers in East Anglia; by the end of 1941, boosted by part-time workers, this rose as high as 30 000. The number of workers in the Cambridge plant itself almost tripled to 3000, though only a quarter were men compared to almost two-thirds in 1939. If part-time outworkers are included, Pye employed some 10 per cent of the 250 000 British men and women who were making radar and other electronic equipment during the war.

The board had already given Pearl the job of advising on matters where 'owing to lack of information, the management might make mistakes in relation to the work-people'. The demands of war were forcing C.O. to pay more systematic attention to his workforce, and the new personnel and welfare department supervised by his sister continued its work after the war. Pearl also took control of the outworkers. Helped by Queenie Culverhouse, her works forewoman from the Invicta plant in London, she turned this unskilled labour into what Pye called its 'Village Industries'. Some relatively skilled help was available, for example from boys at the Cambridge College of Arts and Technology, who had their own benches and tools, but most of the new workforce were women who turned Women's Institutes, village halls, and even their own kitchens into workshops. Eddie and his wife Stella set up a home assembly plant in the outbuildings of their house at Sawston. 'The programme could not be described as economic', a post-war Pye study commented, 'but it was effective.' It was possible because much of the work, such as assembling 12- and 6-pin plugs for the No. 19 set, was simple and needed little equipment beyond a hammer to tap in the pins (even Eddie's 4-year-old daughter Rethna sometimes took part).

Figure 5.5   Stella Stanley (Eddie's wife) outside Sawston outworkers' building with an armful of items to be assembled in No. 19 sets

Vans travelled the countryside to collect completed work and deliver fresh materials.

The scheme was such a success that MAP appointed C.O. its regional production chairman. The Ministry of Production sent a team to Cambridge to study the Pye experiment, and made it a model for

Figure 5.6   Inside the building used for outworking at Sawston

schemes in other parts of the country. The Village Industries contributed to the easy atmosphere at Pye, though on one early morning tour of the plant C.O. did come across a crate of Communist Party leaflets attacking the war (the British Communists opposed the war until Hitler invaded the Soviet Union in 1941). C.O. made a broadcast over the factory loudspeakers challenging the person who brought the leaflets to meet him in the canteen for a fight, but no one turned up. There was also some union unrest, and at one rally outside the University Arms Hotel shop stewards accused C.O. of asking too much from his workforce. But when a speaker attacked him for driving in a large car, which he usually did not do in the war years, a Pye worker shouted back that they should remember the boss had arrived in England 'with barely shoes on his feet'. Union membership at Pye was strong, but seldom militant, and most disputes were between rival unions. Pye was by far Cambridge's biggest employer and competition to work there was strong. Those lucky enough to be chosen expected the job to be for life. 'Once in Pye you regarded it as a family', said Jim Langford, who joined before the war as a 17-year-old apprentice. It was how C.O. saw his company too.

The increase in the workforce took place in spite of the near disappearance of Pye's pre-war bread and butter, the domestic radio set. Military demands on the electronic industry brought about a steady collapse in the production of sets for civilian use. Pye's home radio sales fell by 40 per cent in 1940–1941, and by a further 75 per cent the following year. At the beginning of the war this was partly compensated by a 130 per cent increase in exports which the government encouraged, but these also dropped sharply in later years. The only relief came in 1944 when the government authorised a Utility set whose production was shared among the manufacturers. Because home radio broadcasts were so important to the war effort the BBC wanted half a million of the new sets, but the government allowed only 250 000, and the quota given to Pye and Invicta was just 20 000. Ugly, with a cheap varnished wood case, the Utility did not sell well, and can still be found, unloved and unopened, in its original box at antique radio dealers.

The government spent £50 million on radar equipment each year, and by 1944 the radio industry was two and a half times its pre-war size. C.O. was wary about depending so much on government orders; determination to guard his freedom was one reason he refused to move to a shadow factory. It also explained his initial reluctance to accept grants and loans from the Ministry of Aircraft Production and the Ministry of Supply to expand the plant at Cambridge, though he did accept some government finance for items of capital expenditure. These included a better canteen, an assembly shop and store rooms, while MAP ordered him to put in up-to-date fire equipment, better lighting and a new telephone system. But he remained cautious of government money, and in his 1943 note to fellow directors explained that Pye had only allowed the government to assist on certain expenditures. The amount involved, he added, 'was considerably less than most firms have received and it represents such a small proportion of our total equipment that at no time does it warrant any government interference'.

He had to accept some regulations even if he had no intention of applying them as their authors intended. Stafford Cripps, the Minister of Aircraft Production, who would become Chancellor of the Exchequer in the post-war Labour government, decreed that all companies engaged in war work should have Joint Consultative and Advisory Committees made up of an equal number of workers and managers, their aim being to ensure trouble-free production through participation in decision-making. C.O. admitted the Committee was a useful 'safety-valve' and might educate what he called 'senior work people', but it was not an idea likely to thrive at Pye, and he arranged for Sam Carn, whom he trusted, to be secretary of Pye's JPC. The Council held its meetings, Carn wrote up the minutes in a book which was then put away and forgotten, and that was C.O.'s version of a Joint Production Council. Several months later Stafford Cripps arrived unannounced at Pye to inspect his brainchild. He refused to see C.O., but asked instead for Carn as the Council's secretary. Carn showed Cripps the minute book, which so impressed the Minister that he took it away with him. When he sent it back he attached a letter to C.O. thanking him for the exemplary

Figure 5.7 Listening to Winston Churchill on the wireless at the pub

Figure 5.8 The public listening to Winston Churchill broadcasting

way in which Pye was carrying out his directive (it is not hard to see why Harold Nicolson called Cripps 'a man of great innocence').

How much, and when, the government paid for the military equipment it ordered was a frequent source of conflict. Wartime profits were kept in check by a 100 per cent excess profits tax, but Pye's profits in the six financial years starting in April 1939 averaged out at an annual £101 500, strikingly better than the average annual profits of £80 700 for the previous 6 years of peace. Wartime profits also avoided the peaks and troughs of the pre-war years which ranged from a low £51 000 in 1934 to a record high of £123 000 in 1937, reason enough for Sir Thomas Polson to boast at the July 1944 annual general meeting that Pye's 'unique prosperity' since 1929 had 'suffered no impairment despite the difficult [wartime] conditions'.

From the early years of the war the board minutes regularly refer to the impossibility of compiling accurate trading figures 'owing to the many difficulties and delays which were involved in Government Contract Work'. There was no improvement by 1942, when two-thirds of Pye's estimated turnover was represented by goods produced and despatched but for which the prices had not yet been agreed with the ministries. This situation, in which equipment was despatched, but no invoices issued because ministries had not yet set the prices, continued until the end of the war. Calculating the company's financial position was further complicated by C.O.'s wartime business style. Fred Keys did not much worry about prices because the Ministry of Supply usually settled for a cost-plus method that Pye found easy to manage, though C.O. objected that it gave insufficient reward to efficient companies. A far worse problem for Keys was 'keeping track of time spent on work that was either not officially requested or ... where the source of the instructions had changed or was unclear'. When someone like John Cockcroft visited Pye and made suggestions for new work this might be covered by an official order, or it might be done just because C.O. and B.J. Edwards thought it was interesting and important. 'I'll put a man on to it', Donald Jackson would hear C.O. say, though he 'never had a clue half the time who he'd put onto it'. Keys often found himself having to 'sort out payment for work for which there was no proper paperwork or authority'.

The announced profit figures for the war years can therefore only be considered approximate, a point to which C.O. gave a new twist in his notes for the board in 1943. The profits shown for 1941–1943 were 'far below the real profits earned' because management had to estimate figures for equipment that was still not priced. There was an advantage in this. If the 'large balance outstanding' continued up to the end of the war, as he expected, it would serve as a useful reserve when military orders ran down and Pye took losses on stocks the government obliged it to keep.

CAMBRIDGE **PYE** RADIO . . . .

Figure 5.9   Front cover of a leaflet detailing products made for export, 1941

However patriotic, and however dedicated to the manufacture of the weapons of war, C.O. never forgot about the coming post-war challenge. The government wanted the electronics industry to concentrate all its efforts on war production, but the time came when C.O. and his engineers felt they had done their bit for the government and should now prepare for the changeover to peace. As early as 1943 C.O. gave Pearl the job of planning Pye's peacetime radio production. He had already decided that for the first six months after the war the factory would turn out only pre-war models; Pearl was to decide on new sets to be put on the market after that. C.O. expected that wartime scientific work would create new lines of both industrial and scientific equipment, but they would cost money to develop. Pye would therefore sell fewer radio models than in the past because radio would 'have to carry the normal overheads of [the rest of] the business'. For the same reason Invicta would have to slash its overheads by producing 'a simple radio and nothing else'. His plans took on enough detail to include which components Pye would make for itself rather than buy in; all executives were ordered to attend monthly meetings on the preparations for peace.

One man who needed no encouragement to think about the future was B.J. Edwards. At the beginning of 1941 he had produced a paper on the design for a post-war television using the EF50 valve. Three years later he took his ideas further in a more general paper delivered to the Institution of Electrical Engineers. It was a call for Britain to build on the lead given it by a new military technology that would make possible both the mass-production of TV sets and the rapid introduction of an advanced new system based on as many as 800 lines (compared to the pre-war 405-line standard). In 1943 members of the Pye radar team began work, discreetly, on Pye's first post-war TV. Development of the aerial was done in the privacy of a wooden hut on the sports field. Any equipment obviously for use in a television set, such as cathode ray tubes suitable for 405-line receivers, was moved out of sight whenever government inspectors visited the plant. By such means C.O. made sure that when television transmissions began again Pye, alone among the manufacturers, would have an entirely new set on sale in the shops.

The war years brought about the complete identification of C.O. with his company. He had not shown great interest in the routine of peacetime production and in the 1930s spent as much time in London as in Cambridge. War challenged him, and he was in his element in a wartime factory where so little was routine and so much unpredictable. These years also brought him closer to the scientists and engineers who always interested him because they, too, chased after original ideas, and it was one of them, Dennis Fuller, who felt that from this time C.O. '*was* Pye, good and bad'.

The war years were difficult. People were usually tired: the middle-aged men the young Anita Sturrock found herself working among in the research laboratories were in fact half-exhausted 20 and 30 year olds. The radar engineer Leslie Germany, who developed GCI, worked from nine in the morning until nine at night on weekdays, and until

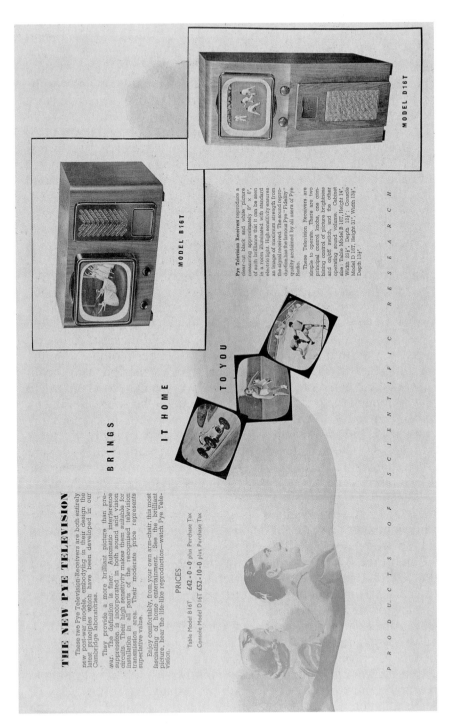

Figure 5.10   Inside of a leaflet, 1947, on the first post-war television receiver, model B16T, made by Pye in 1946

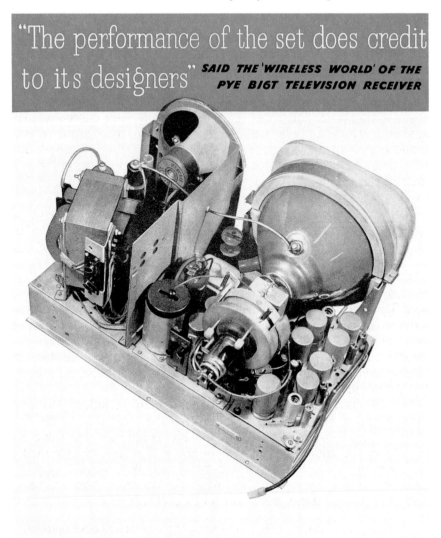

Figure 5.11   Chassis of Pye model B16T television receiver

five pm at weekends. He was also a member of the Pye section of the St John's Ambulance Brigade, sleeping one night a week in the factory surgery, and fire-watching another. Because so much of the factory's work was secret, fire-watching had to be done by senior staff. Fred Keys slept in his office twice a week, and attended two weekly Home Guard drills and a parade on Sunday mornings. Most people did not expect, and certainly did not get, pay for overtime. For the first 2 years of the war, and particularly after the gloom that fell on the factory after Dunkirk, they worried about the prospect of a German invasion. The mood changed with the tide of war. News of the D-Day landing was announced over the factory loudspeakers, and when VE Day came most people slept. 'It wasn't tiredness by then', Anita Sturrock thought. 'It was relief.'

Many at Pye remembered the war with unmixed satisfaction. Bill Pannell, the young designer of the No. 19 set, considered it 'the most exciting time that any of us had had'. If anyone in Pye was dissatisfied it was C.O., who was hungry for public recognition of what he and his company had achieved. In early June 1945, a month after the end of the war in Europe, he used disclosures about radar in an American magazine as a pretext for a press conference in Cambridge at which he described the still secret operation of Air Interception (AI) and Ground Control Interception (GCI) radar. Radar, he said, had made it possible for the RAF to defeat the German air attack, and pointed out that Pye had worked on the development of both systems almost from the start.

The *Times* reported his remarks at some length, but this small step towards the recognition he wanted for Pye was forgotten in August when Stafford Cripps, now President of the Board of Trade, paid tribute to the inventors of radar in the House of Commons. He praised the contributions of the Armed Services, the ministries, the government research establishments and the universities, but said nothing about the industry that not only made the radar sets but contributed so much to their development.

C.O. was furious, and organised the leaders of the industry to lobby Cripps to make a second speech repairing the omissions of the first. Cripps gave in almost at once and, unable to return to the subject in the Commons, addressed a lunch of the Radio Industry Council at Claridges on 31 August 1945. Robert Watson-Watt, always generous in his recognition of industry's contribution to radar, wrote Cripps a draft, and the resulting speech was an elegant tribute to

> the other great partner in this team-work of research and development, the scientists in the laboratories of the industrial firms who by their ingenuity and resourcefulness first turned the rough models, and even in some cases the ideas, of the government scientists into devices which could be quickly and economically produced in the factories and which, in the exacting conditions of military service, were capable of giving good and continuing performance.

Cripps stressed that where radar was concerned there was no difference between industry's laboratories and the government's: 'we treated all as partners in a single fellowship of science'. He outlined the history of radar with its constant search for more powerful and more sensitive valves to produce the ever shorter wavelengths on which the precision of the radar set depended, and alluded to, though he did not name, the

EF50 and the Pye strip. Cripps singled out only 12 firms for special mention, putting them in three groups. Pye was honoured by being grouped with Metro-Vickers and Cossor, the two veteran suppliers of electronic equipment to the Services, which had built Britain's first radar system Chain Home, and with the large General Electric Company, co-inventor of the cavity magnetron. All four were praised for their 'outstanding contribution' to 'scientific research in the laboratories'. The remaining eight, among them Ferranti, Ekco and EMI, were commended chiefly for their development work.

Cripps ended with a survey of radar's peacetime prospects. He was 'most anxious', he said, that other countries should understand 'what a magnificent industry we have built up during the war so that they will flock to you with their orders'. It was nevertheless clear that Cripps, like most of Britain's intellectual and governing establishment, put science on an altogether higher plane than industrial production, and that he was doing Pye a special honour by placing it in the category of firms remarkable for their scientific achievement. C.O. knew all about Whitehall's condescending attitude to industry, and had himself sometimes behaved as though leading his factory into war was not honour enough. On at least two occasions he expected the offer of some important job connected with the war effort (this was aside from his temporary work for Sir Robert Renwick smoothing the supply of components to the radar and radio manufacturers). The Pye minutes also record that in August 1942 C.O. thought he might get an appointment in the Armed Forces. His directors 'pressed [him] not to take this decision finally before some little time had elapsed'. C.O. agreed, but warned 'he might have to ask for a leave of absence if he went on active service'.

It is hard to tell whether he was serious, or just teasing his colleagues to make them recognise how completely they depended on him. Fred Keys attended enough of C.O.'s meetings with high-ranking officers and senior civil servants to know how many of them 'hated' him, not least, Keys thought, because he was so often right and they wrong. 'You could usually tell who did and who didn't like C.O. It was always the . . . unconventional and odd types [from the Services and government establishments] that used to get on with him.'

C.O. believed he could ignore enemies with impunity, which was not a recipe for winning official favour, let alone an appointment in the Armed Forces. This was brought home to him by the publication of the 1943 honours list, in which he was awarded an OBE. It was an honourable decoration, but by then Pye had produced Britain's first airborne radars and the unrivalled No. 19 radio set, and also made its vital contribution to the development of the proximity fuze. 'It's a bloody disgrace', C.O. said when he heard the news, and asked Norman Twemlow what he would do. 'Tell them to stuff it', said Twemlow, who enjoyed being a blunt Mancunian, though he later advised C.O. to accept

Figure 5.12

Figure 5.13

the medal in the expectation of something better later. C.O. took the advice, but Pye was again meagrely rewarded in the post-war honours list of 1946. B.J. Edwards got a lowly MBE; Harmer, who so unusually for Pye got on well with top brass, an OBE. C.O. was raised just one rank to CBE.

Ian Orr-Ewing, the young Air Force officer who took samples of the EF50 from Cambridge to London at the start of the war, said later (when he was himself a life peer) that C.O. deserved a peerage just for giving Britain that valve. It was, he thought, a 'mark of [C.O.'s] perversity that he failed to understand the concern, anxiety and irritation his provocative views caused people in high office'. That was true, but it was not the whole story. The awards given C.O. and members of his staff were different gradations of the Order of the British Empire, and in 1946 there was still a British Empire and, in Whitehall, a great government machine that was self-consciously imperial. Industry and commerce occupied a subordinate place in the hierarchy of the Empire; they were the figures representing human toil that appeared at the base of monuments to the monarchs, marshals and proconsuls who were still the Empire's deities.

Under this old establishment, industry might be inferior, but it was at least allowed its sense of honour. The Labour government elected in July 1945 questioned the very morality of business committed to the pursuit of profit. And in men like Stafford Cripps it would have ambitious, able ministers who believed their experience in the war had proved the advantage of planning over the unorganised commercial competition they despised. The war with its triumphs and disappointments was over; it was time for C.O. to enter the new battlefield of a Labour-governed Britain.

Sources

COS files: 1/8/2, 2/3, 5/1 (outworkers).
COS/VDS1/3 (Velma's dispute with ARP).
Simoco boxes: Box 1 (board meeting agendas 1940–1946).
Pye main board minutes.
Interviews: John Anderson, Brian Beer, Audrey Darkin, Dennis Fuller, Donald Jackson, Fred Keys, Jim Langford, Marjorie McCarthy, Patsy Morck, Bill Pannell, Anne Powell, Nancy Sturrock, Peter Threlfall.
Michael Bell interview: Les Germany.
Books: Bell, *ibid.*; Briggs, *ibid.*, Vol. III; Crowther, *Science at War;* Geddes and Bussey, *ibid.*

Chapter 6

Boom and bureaucrats

The peace for which C.O. Stanley had stealthily prepared since 1943 brought even greater opportunities than he foresaw. Britain's spending on household durables quickly reached, and then overtook, the level of 1938. High employment and growing industrial production would by 1955 make the real average weekly wage a third higher than in the last years before the war. With Germany and Japan in ruins, and American industry more interested in its home market, Britain's manufacturers had by 1950 captured one-fifth of a booming world trade in exports.

After winning the war the British expected a better life, which was most easily measured by new houses and more consumer goods. It was said of C.O. that he could 'sell diamonds from the back of a jaunting-car' (the poor Irishman's pony and trap), but it took no rare talent to sell radio and television sets in early post-war Britain. The government's wartime investment in the electronics industry had increased its productive capacity, and by 1947 radio sales were back at pre-war levels.

The first new Pye sets, prepared under Pearl's supervision, were better than the company's pre-war models but still looked old-fashioned. C.O. had no more taste for modern design than Velma, who furnished their houses with antiques, and he was unimpressed when the Festival of Britain brought modern design to the British in 1951, telling his staff 'I don't want any of that Festival stuff here'. Nevertheless he did hire the well-known designer Robin Day to make Pye radio cabinets as up-to-date as the advanced multi-waveband receivers inside them. Radio was still an exciting medium, irreplaceable as a source of entertainment and information. When C.O. gave a helpful station master one of Pye's first post-war sets he received a note of almost rapturous thanks. The 'greatly treasured gift', the railwayman wrote, had 'taken up a very prominent position in our dining room' to which it added 'splendour',

Figure 6.1   Front cover of a personalized letter from C.O. Stanley to commercial dealers reassuring them that their supplies would be restored as soon as the war allowed, 1944

and he had 'sat up until the early hours of this morning enjoying the various programmes, excluding the Third'.

C.O. scored another coup when Pye brought out the Black Box record player in 1953 under licence from CBS. He was so taken with

Figure 6.2   Listening to the new 'Baby Q'. Illustration from a Pye 'Circle' magazine, 1947

its handsome lines and quality of sound reproduction that he demonstrated it himself to dealers. This endeared him to the shopkeepers Pye depended on for sales, for they were not used to seeing a major manufacturer appear in front of them as though he were a regular salesman. But it was television that held his imagination. Annual radio production peaked at 1.9 million in 1947 and fell away to 1.6 million by 1955. In 1946 just a few thousand television sets were made, but by 1955 Britain would have a total of 4.5 million. No one was better prepared than C.O. to attack this market. He got round the problem of testing the B16T, his first post-war set, before the BBC resumed television transmission in 1946 by making his own, technically illegal, broadcasts. New sets were installed in the Cambridge houses of senior employees and for an hour each evening Pye transmitted little TV shows in which secretaries read the news and Eddie and other amateur musicians gave recitals. Thanks to such devious preparations the B16T, price £35 plus £7 17s 3d purchase tax, could be launched on the market in April 1946, 2 months before the BBC resumed its television service.

Figure 6.3   Pye Chinese lacquered 'Black Box' record player, 1953. Contrary to its name most 'Black Boxes' were in a mahogany finish

Long brooded over by B.J. Edwards and designed by the radar team led by Ted Cope, it was the first British set on the market with the EF50 valve that Pye had planned to use in 1939. More compact than earlier sets and easier to service, the B16T was the most advanced television Britain had seen, while the first post-war sets of other British manufacturers were old models with or without cosmetic changes. *Wireless World* called it 'very considerably superior to pre-war models', but it was only the first in a series of Pye innovations. Its successor, the B18T, was the first television that did not need a mains transformer and, because of the consequent reduction in weight, arguably the world's first portable set (credit for this went chiefly to Pye's refugee from Germany, Ladislav Lax, who had experimented with the technology while working for Telefunken before the war).

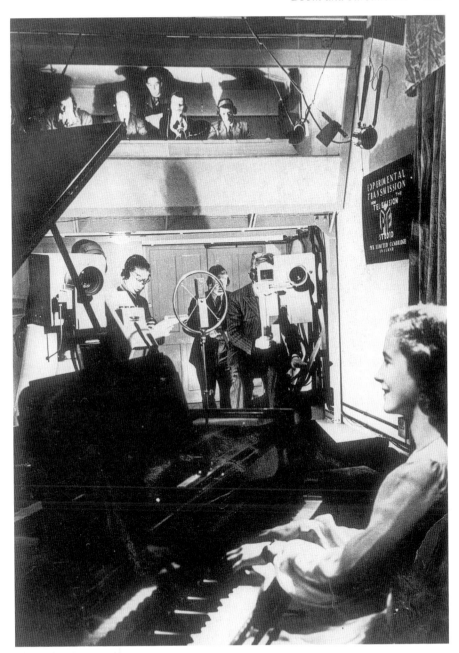

Figure 6.4   Pye's experimental television studio, Cambridge, 1946

Robin Day's changes to the appearance of Pye radios affected television sets too. The V4, released in 1953, not only looked recognisably modern but was the first British set with a black screen to give a better daytime picture. It also came with Automatic Picture Control, devised by Leslie Germany to prevent outside interference such as a plane flying overhead breaking up the picture – a chronic problem in early television sets. The VT4, seen at the Radio Show the next year, was the first tuneable 13-channel set that could receive BBC transmissions throughout the country as well as the new ITV transmissions due to begin in 1955 (up to that time sets were pretuned to receive transmissions of the area they were to be used in).

C.O. distanced himself from the rest of the industry in other ways too, in 1946 resigning from the council of the British Radio and Electronic Equipment Manufacturers Association (BREMA) after a disagreement over television development. Fearing damage to sales of old stock BREMA did not want to publicise new sets that could be adapted to receive the new channels created by the BBC's expansion of transmissions. C.O. typically chose to pursue this quarrel in the Pye annual reports which he turned into a bully pulpit from which he mocked and hectored anyone who obstructed his progress. After Pye's introduction of the tuneable VT4 he used the annual report to pour scorn on competitors who argued it was cheaper to adapt existing sets to receive new channels. Nonsense, C.O. declared. 'Pye, usually the first in the field in our industry, has the right answer – the new tuneable television set . . . for the shape of things to come.' Relations with his industry, as with many other groups and individuals, were taking on a predictable, switchback pattern. 'He would fight you one moment', a BREMA official recalled, 'and be very generous the next . . . [He was] a spur to progress, but with a one-track mind.'

Life in Pye's post-war laboratories remained as stimulating as it had been during the war. For Jo Fletcher, working on television under Cope, it was 'a constant daily "high" . . . an atmosphere where innovation and personal initiative were respected and admired'. C.O. had boasted before the war that 'where Pye is today the rest of the industry will follow tomorrow', and his engineers were making good that claim. Much of the credit went to B.J. Edwards, who worked from a glass-fronted office in the Upper Research Laboratory that allowed sight of everything around him. He was a difficult taskmaster who enjoyed playing slightly cruel practical jokes, and sacked people for trivial reasons only to expect then back on the job the next day. But his wartime achievements gave him glamour, and those with good nerves found his unpredictability exciting. Like C.O. he kept ahead of the game, steering his research team towards projects that only became practicable years later. Among the ideas he threw out were a fax machine, an automatic photocopier and

Figure 6.5   A graphic depiction of how the V4 receiver held the television picture rock steady

Figure 6.6   Production line for the Pye VT4 television receiver, Pye factory, Christmas 1954

Figure 6.7   Pye VT4 television receiver on a complementary stand, 1954

Figure 6.8   The National Radio Show, Earl's Court, London, 1954

Figure 6.9   Upper Research Laboratory at the Pye factory, Cambridge, as seen from the office of B.J. Edwards, c.1947

Videosonic, a pioneering method of combining the sound and picture signals in a television set instead of transmitting them separately. Pye tried to sell it to the Post Office and the BBC in 1946, but it was not practicable until transistors replaced valves.

C.O. contributed to the excitement by taking every opportunity to show off the still young medium of television in the most seductive light, and in the spring of 1949 helped the BBC televise the entire Oxford and Cambridge Boat Race. It was the first time two moving objects (the crews) had been televised from a third (the following launch), and over a distance of 4 miles. Pye provided all the equipment: the launch *Consuta* rigged out as a mobile unit, eight cameras developed by Cathodeon and several miles of land lines along the river banks. There were two transmitters that fitted into boxes no bigger than suitcases and at the finish the mobile Outside Broadcast Control Unit that Pye had only just delivered to the BBC. The result was a double triumph – for Cambridge over Oxford, and for C.O. Stanley over *his* competitors. It was also a source of pride that he had pulled off a coup so much to the taste of that passionate oarsman, his father.

The Boat Race broadcast showed how far C.O.'s interest in television was moving beyond the simple TV set. With Cathodeon's manufacture

*Figure 6.10   Pye television equipment on board the launch used for the first television
broadcast of the Oxford and Cambridge Boat Race, 1949*

of camera tubes he had broken into the market for television cameras, and later strengthened his hand by co-operating on tube development with America's General Precision Laboratories. B.J. Edwards encouraged him to move in new directions, and it was Edwards who sent Ted Cope and Leslie Germany to consult with Pye's old alumnus Peter Goldmark, who was working on a colour system for CBS. C.O. allowed Edward's £25 000 (almost 10 per cent of Pye's 1948–1949 profits) to develop colour television 'on the Goldmark principle' and demonstrate it at the 1949 Radiolympia, though only as a medical and industrial aid and not, as Goldmark intended, for domestic receivers. Described by C.O. as 'the first really successful colour television system in England' it was tried out in two London hospitals, and demonstrated by Pye in the United States and Europe.

It was not long before C.O. was thinking of colour for the home, and the Coronation of Elizabeth II in June 1953 was his chance to publicise its possibilities. BBC cameras showed the procession and the ceremony inside Westminster Abbey in black and white, but C.O. got permission to put one colour camera on the roof of the Ministry of Health overlooking Parliament Square and another in Whitehall. Pictures were transmitted by UHF link to receiving aerials at the Great Ormond Street Hospital for Sick Children, at the Fleet Street offices of the *Daily Express*, and at Treetops, the Park Lane flat of Sir Robert Renwick, the wartime friend who had become C.O.'s closest business ally.

Pye's colour pictures showed only part of the procession, and because they were shown on large studio-size monitors lacked sharpness, but they were the only colour pictures anyone in Britain saw that day. C.O. and Renwick watched at Treetops with their families and some mutual friends. Celebratory drinking began early in the day, with many toasts to C.O. as the creator of this marvel. Guests were so thrilled to see the brilliance of the Blues and Royals riding beside the Queen's golden coach that they spent much of the time telephoning friends to boast they were watching the ceremony in colour. C.O.'s imagination had paid off and out of excitement and, perhaps relief, he overdid it. Willie Wakefield, the young Pye engineer looking after the screen at Treetops, stepped out of the Renwicks' drawing room to find his employer standing unsteadily in the corridor. C.O. called out for help, and the young man guided him to a bathroom where C.O. was thoroughly sick. Wakefield discreetly got him to bed and brought him a cup of tea.

Britain watched the Coronation on 2 million television sets, many of them made by Pye or its chief competitor, Bush. Pye had taken the lead with the launch of the B16T in 1946, when C.O. estimated that 5000 of Britain's almost 20 000 pre-war sets were still in working order. By the early 1950s, when Bush started to draw even in sales, Pye was making 1200 sets a week and C.O. expected profits from television sales to cover all the company's normal overheads, turning income

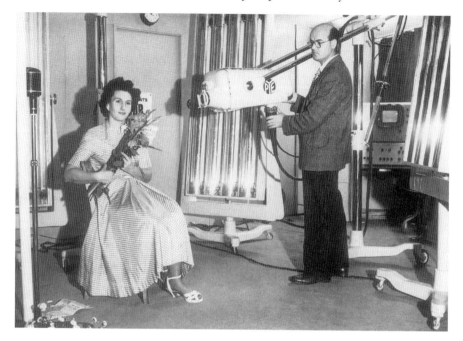

Figure 6.11   *Televising a bouquet of flowers in colour on the Pye stand at Radiolympia, 1949*

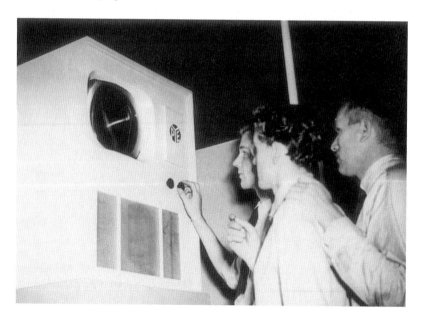

Figure 6.12   *Visitors to Radiolympia getting their first glimpse of colour television, 1949*

Figure 6.13   Press pass for Pye engineer involved in their limited colour transmission of the Coronation, 1953

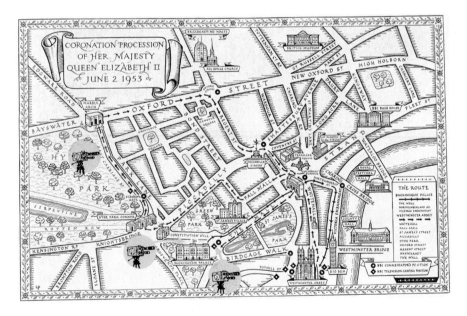

Figure 6.14   Map of the Coronation route, 1953. Pye's television cameras (black and white) played a highly important part in this memorable outside broadcast

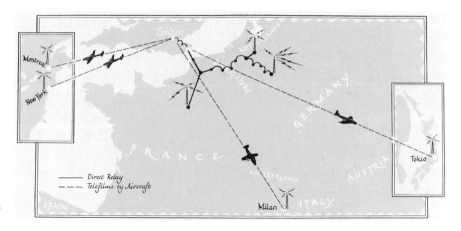

Figure 6.15   Map showing worldwide coverage of the televised Coronation, 1953

Figure 6.16   Family watching the Coronation in black and white on a Pye V4 television receiver, 1953

Figure 6.17   Family watching colour television on a Pye experimental receiver, 1954. This venture follows the successful limited coverage of the Coronation in colour by Pye that was relayed live the previous year to the Great Ormond Street Hospital for Sick Children and certain other venues

from other parts of the business into net profit. The success was all the more striking because, judged by its capital, it was still a minor player in the industry. A 1949 government survey of the capitalisation of firms making television sets showed GEC, with issued capital of £9.8 million, and HMV/EMI (£3.4 million) to be the giants. Bush had £1.2 million, while bottom of the list McMichael Radio had £262 000, only just below Pye with £366 000.

By 1955 Pye's profits were £2.2 million compared to £263 000 in 1946, while its workforce had grown to 10 000. This transformation from ambitious but provincial firm into national player made itself felt on factory life. In 1950 Charles Harmer told fellow directors that 'he had been in touch with Mr A. Duxberry, a well-known teacher of elocution, with the object of arranging a course of lectures for senior executives'. In 1954 C.O. decided that his company needed an office building that reflected its new status and asked the architect Sir William Holford, known for his work in the City and around St Paul's, to submit a design. C.O. rejected Holford's first effort as 'very grandiose', and asked for 'a straightforward two-storey building' for the site opposite the Pye factory on St Andrew's Road. When the revised design was priced at £300 000 he abandoned the professor. A little later, though, he bought for himself and Velma a Regency house in Little Shelford outside Cambridge. Sainsfoins was a country gentleman's handsome estate excellent for business entertaining, but also where he would raise pedigree cattle while Velma took command of the gardens.

Peter Threlfall, Pye's export manager after 1946, remembered the early post-war period as 'an endless boom, [when] no one worried'. But C.O. worried, not because he thought his markets might disappear, but because he feared his freedom to exploit them would be constrained by governments indifferent to what he saw as Pye's and the country's true interests.

He blamed most of the things he disliked on the 'socialism' and bureaucracy of the Labour government elected in 1945. It was hard for him (and many others, too) to accept that Britain's chief problem in 1945 was neither socialists nor bureaucrats but the appalling state of the country's finances. War had consumed the nation's assets and Britain was left with the biggest external debt in its history. This 'financial Dunkirk', as the economist J.M. Keynes called it, determined the conditions under which successive British governments strove to achieve economic growth without toppling the country into bankruptcy. Labour's construction of the Welfare State added to the strain; so did the commitment of both main political parties to Britain as a world power. The result was a weakened pound and limited freedom of action for those who guided the economy, and a dreary pattern developed of excess demand triggering financial crisis which triggered restrictions on the demand that had set the process off. This stop–go economy, as it came to be called, was particularly irksome to manufacturers of consumer goods, for these were an obvious target for any Chancellor of the Exchequer who needed to cut back demand.

It was easy for Labour politicians to see television as an unnecessary luxury in a country recovering from war. Sets were not only expensive but also useless in those parts of the country where BBC transmissions did not yet reach, and when Labour's first Chancellor Hugh Dalton introduced purchase tax in his first budget it fell heavily on radios, radiograms and television sets. The following year he doubled the tax, though his successor, Stafford Cripps, fearing deflation, abandoned plans for a further increase. In 1952 the new Conservative government added a further tool of demand control by altering terms of the hire purchase agreements that accounted for a quarter of radio and TV sales, setting a minimum deposit of 33.3 per cent and a repayment period of 18 months.

Manufacturers protested, and none more persistently than C.O., who in 1948 tried unsuccessfully to persuade the Chancellor to receive a deputation to discuss purchase tax. Pye's annual report for 1949–1950 was illustrated with a drawing of Gulliver struggling to escape from ropes pegged into the ground by Lilliputian bureaucrats and politicians. One rope around Gulliver's legs was labelled 'government majority', another 'monopoly broadcasting', but the rope round Gulliver's chest was purchase tax, 'probably the worst weapon the little men ever conceived.' In a letter to the *Financial Times* C.O. claimed the tax disrupted both trade and production, explaining that the public would not buy a television in the run-up to a Budget if they thought the Chancellor might reduce purchase tax. This, in turn, caused shops to cut back orders and left manufacturers' production schedules in disarray.

Another consequence of Britain's precarious finances was the continuation of controls introduced during the war. These affected labour, investment, and the supply of almost all the main industrial raw materials including hardwood for television and radio cabinets, and glass for cathode ray tubes. Shortages at Pye were serious enough for C.O. to chair a weekly production committee to chase up missing materials, and he got on the telephone himself to charm and bully suppliers into releasing stocks. When rearmament for the Korean War led to new shortages in 1950 C.O. had to send a sales representative to live in Birmingham to make sure Pye got even small batches of sheet metal needed for television chassis. C.O. was not alone in feeling that the Labour government, and not least the powerful Stafford Cripps, whom Tories mocked as a 'teetotal totalitarian', believed in austerity for its own sake.

At moments C.O. thought Labour might extend nationalisation throughout British industry, a fear encouraged by the experience of

Figure 6.18   *"... the giant of private enterprise, having at last shaken one leg free, raises its virile frame from the shackles of the midget planners that have fettered industrial endeavour during the last five years." Pye Annual report, 1950*

his friend Robert Renwick, whose family firm, the County of London Electric Supply Company, was nationalised in 1948. Renwick became an active Tory supporter, explaining later, 'if I have a scar in my heart it is because [the Labour government] took my industry away and made a mess of it'. C.O. supported Renwick in many of his political activities but there was an important difference between them. For C.O., Labour was just a particularly dangerous species of a troublesome genus; he did not trust politicians of any party to let him run his company as he thought best.

Fear that his industry might be nationalised preoccupied him in 1948 when he was planning to set up a new subsidiary to make radios and television sets. The government wanted Pye to observe the spirit of the 1945 Distribution of Industry Act and pick a site in a distressed part of England. C.O.'s justification for not doing so was that government controls 'had created the biggest "spiv operation" imaginable'. Instead he picked Ulster, hoping that if Pye itself was ever threatened with state ownership, the subsidiary could become a company controlled in Northern Ireland and beyond the nationalisers' reach.

His precautions were unnecessary because Labour was itself divided on nationalisation. Cripps thought it pointless, believing the war had shown that government could control the economy by planning alone. This was scarcely more agreeable to C.O. and in 1947 alarm bells rang in Cambridge when the Board of Trade began a census of the industry. C.O. ridiculed it as a waste of time, but pointed out to the Pye board that firms failing to comply 'might well find themselves in difficulties later [over] supplies of material [allocated by the government]'. Judging by the reply to the Board of Trade from Pye's white goods subsidiary L.G. Hawkins, C.O.'s tactic was prevarication. A long, obstructive letter explained that Hawkins had neither 'clerical staff nor records available to arrive at [the] detailed information' the census demanded. 'We tremble to think of the time spent [on it] by members of our staff (this writer alone has spent weeks).' The clinching argument was that the hours wasted filling in the questionnaire would damage all-important exports. Whitehall was not impressed and reminded Pye that it was conducting the census under Defence Regulations and there was a statutory obligation to reply.

The Board of Trade enquiry was supposed to contribute to what Cripps called '[matching] our resources against our needs so the main features of the economy may be worked out for the benefit of the community as a whole'. By the end of his time as Chancellor, though, he was talking less of planning and more about the budget as 'the most powerful tool to influence economic policy . . . available to government'. Given the budget's manipulation of purchase tax and the terms of hire purchase this, too, brought C.O. little comfort. Nor did Labour's habit of disparaging business and businessmen. The left-wing publisher Victor Gollancz put his finger on the contradiction in Labour rhetoric: Cripps would urge businessmen to cut costs and increase production in the national interest, but in the same speech, 'these same men, in their aspect as profit makers . . . are sneered at and talked about as if they were criminals'.

Small wonder C.O. often reacted to the government with the anger of hurt pride. He had taken to business as a bird to flight; it was the element in which he lived. Brought up within a traditional morality by parents he admired, and not given to self-reflection, he did not accept that others might question the basis of his life. The war had accustomed him to think of Pye and patriotism as synonymous, and the achievements of the post-war years strengthened his natural self-assurance. How could he not fight back as he did in Pye's defiant 1949–1950 annual report? Others helped prepare these documents, but only C.O. spoke from their pages, and his voice rang clear in the assertion that the time was gone 'when the industrialist could get on with his work and leave politics to the politicians'. Some businesses and trade associations had failed to understand this and had 'so bent their backs to the whip of the planners' notions that . . . they will never be able to stand straight again'. Pye was not like that. 'When politics interferes with us, our workpeople, and our shareholders . . . we must support those political opinions which support the principles of incentive and of private endeavour.'

The chance to get rid of the Labour government in the 1950 elections inspired him to launch the Free Enterprise Circle made up of employees of Pye companies spread over some 20 parliamentary constituencies. Circle members made speeches at Conservative rallies and heckled at Labour ones, and also distributed the Circle's own leaflets and stickers. At a meeting at the Cambridge Guildhall on 21 February, C.O. re-stated his right as a businessman to speak out about politics: after five years of politicians interfering with Britain it was time '*we* interfere with the politicians'. He compared recent remarks by Prime Minister Clement Attlee and the writer and Labour supporter J.B. Priestley to speeches made by Hitler's propagandist Joseph Goebbels, and ridiculed Labour's claim to have brought down the death rate when credit belonged to the private enterprise that produced drugs such as M & B and penicillin. And he mocked the government's devotion to food rationing by describing how when Pye sent sick employees to recuperate at a cottage at Lisselan, his Irish estate, their austerity-conditioned stomachs could not cope with the 'ordinary good food' of County Cork. As for Labour's boast of bringing full employment, post-war conditions were such that if 'we had had a government of the maimed and the blind, the deaf and the dumb, we could not have had unemployment'.

A particular target for his scorn was the Labour MP for Cambridge, Leslie Symonds, whom C.O. accused of raising 'the threat of the Iron Curtain' when he held an election rally outside the Pye factory. Symonds had suggested C.O. was planning to 'pack up [Pye] and go elsewhere' if he disliked the election result. There was a grain of truth in this. Just before the election C.O. took the trouble to have the Pye minutes record the board's 'firm belief in private enterprise and its firm intention, in the event of a Labour Government being returned to power . . . to concentrate on the development of the company's oversea businesses'. Symonds' sin was his inability to understand why C.O. felt as he did. He had set up companies in Australia and Canada in 1948 because of 'the expense brought about by controls, restrictions and frustration' in Labour-governed Britain. He claimed that for the same reason Pye (Ireland) had to make Pye's export model TV sets because it cost too much to make them in Cambridge.

The MP's most grievous mistake was to warn C.O. that the government had the power to 'stop this sort of ratting'. One can feel the fury in C.O.'s counter-blast: Aneurin Bevan had called the Tories 'lower than vermin' during his battle to introduce the National Health Service; now it was 'rats'. Was C.O. 'ratting' when he joined the RAF in World War I? And what was this 'power' that Symonds claimed to have? 'Is this a slave state? . . . You, Mr Symonds, belong to a government which, when somebody has invented something, or inspired something worthwhile, will steal it and call that nationalisation.'

Labour's assaults on his pride and his business explain why he joined Robert Renwick in reviving the Institute of Directors. In 1948 the Institute was a sleepy institution with a few hundred members and little income. Renwick and like-minded friends saw it could be turned into a pressure group to argue industry's case in a hostile world, and carried out a coup to remove the old leadership. Major-General Sir Edward Spears, a distinguished soldier who served as Churchill's wartime liaison officer with General de Gaulle, was elected the new chairman; C.O. and Renwick took seats on the Institute's ruling council along with grandees such as Lord Derby, Oliver Lyttelton and Walter Monckton as well as leaders of ICI, Fairey Aviation, the *Financial Times* and other well-known companies.

A resolution of the new council in September 1949 that might have been written by C.O. attacked 'the constant growth of interference by the government with the freedom of industry and commerce', and urged businessmen to fight back. Spears' early speeches were marked by the same Stanley spirit, notably a rallying call to end the 'ditherings of that ... most sniped at of mortals, the company director'. Predicting 'the road out of socialisation may be very long', Spears warned businessmen they would have to get used to working in a society where 'the public has been taught to regard company directors as the drones of industry'. C.O. took up the theme when he addressed the Institute's conference in the spring of 1950. The Labour government was by then in retreat, and though he still sounded embattled ('We find ourselves attacked ... from every side') he argued that the Institute had been revived for 'some higher motive and bigger principle' than the protection of the business community.

> It is not often understood by the politicians who attack us that we have a very broad moral outlook ... [and that we have] responsibilities to the people who work for us and ... [who] give us credit [as well as to shareholders] ... We must have an ideal [so] that the people who come into this profession shall be fitted to come into it ... a standard of morality ... that makes it worthwhile to lift this profession to the forefront of the great professions of this country.

There was humbug here. C.O. certainly felt responsibility for employees, rather less for some of his creditors and shareholders. Quite genuine, though, was his wish for society to appreciate the contribution to the common good made by business in general, and C.O. Stanley in particular; in other words for others to share the good opinion he had of himself.

A history published by the Institute of Directors on its 80th anniversary gave five men credit for its regeneration as a 'base to fight socialist economics': Spears, Renwick, Oliver Lyttelton, the Earl of Drogheda and C.O. Stanley. In fact he was rather offhand in his dealings with the Institute. Elected to its important policy subcommittee, he seldom bothered to attend its meetings. Spears coaxed him with compliments – 'I can think of no one whom I would prefer to help form . . . policy than yourself' – but C.O.'s attendance got no better. Some years later the journalist Anthony Sampson would describe him as 'one of those implacable non-Englishmen who never succumb to the charm of monopoly or rings'. He might with equal truth have added 'or of clubs or any other of the associations popular with the British ruling classes'. His clubs were usually connected with his sporting interests: the Teddington Hockey Club (he and Rue were keen players in Arks days), the Royal Wimbledon Golf Club, the Royal Thames Yacht Club. He never tried to join the smart London clubs where Robert Renwick was so at home.

He was no more of a committee man than a joiner. The only committees he could be counted on to attend were those he ran himself. The one time he took full part in the work of the Institute of Directors was in 1960 when he bullied his colleagues into setting up an export committee under his leadership. He thought up Export Action Now, a campaign to instil in the public some of the enthusiasm for exporting that C.O. showed at Pye, and his message was that unless Britain exported more it would 'end up as a second or third rate economic power'. He was preaching what he himself had practised since the war. By 1955 Pye was Britain's largest exporter of radios with total exports of more than £2 million, and double that amount by the time Export Action Now was launched.

Once in charge of his own committee he was full of ideas – for a barometer in Piccadilly Circus showing daily export figures; for citations for 'services to exports' in the Honours List; for the political party research groups to turn themselves into export think-tanks. The press took up the campaign but C.O., who did not see the need to canvas anyone else's opinion on tactics, was disappointed in his business colleagues. When the President of the Board of Trade, Reginald Maudling, switched on an illuminated Export Action Now sign in Piccadilly in June 1961, the Institute's director general, Sir Richard Powell, had to apologise to C.O. for being unable to arrange 'an amusing dinner party' after the ceremony because 'everyone was at Ascot'.

C.O. never went to Ascot. As an Irishman who made his money in manufacturing he was not easily accepted at the top of a society whose powers of exclusion seemed to have survived the war intact. Perhaps this partly explained his guarded attitude towards the Conservatives, for instead of welcoming their defeat of Labour in 1951 he sounded a warning. A headline in the Pye annual report for that year declared 'we criticise any Government' and he attacked the Conservatives' first budget for still showing a tendency to class warfare. He detected 'a remnant of Victorian snobbery when to be "in trade" was a bar to social advancement'. 'To *make* something, particularly with profits, still seems to have rather a bad odour...We pray that our new masters have enough businessmen among them and few enough academic theorists to realise that Big Business is not the Big Bad Wolf'. The new government had also missed a chance to start dismantling nationalised industries, that 'monstrous incubus with twin nightmares of ever-rising costs and ever-rising prices'. If ministers still needed proof of the continuing spread of bureaucracy in post-war Britain they should visit Pye's Cambridge neighbour, the East Regional Government, which had swollen into a 'hut-like encampment...employing thousands of people... to rubber-stamp the vast quantity of paperwork to justify its existence'.

No industry could escape the crippling consequences of the nation's invalid finances, but C.O.'s business suffered from a peculiar disadvantage: the further he took Pye into new technologies the more he had to deal with Whitehall. The main source of this conflict was Pye Telecommunications (known within Pye as Telecomm), set up in January 1948 with a £100 000 loan from its parent Pye Ltd. Eddie Stanley was the new company's chairman and C.O. attended only the first meeting of its board, appointing as managing director Harry Woolgar, who had run a wartime section in Pye that provided preproduction units for urgent military projects. A workforce of 30 in a Nissen hut next to the Pye canteen began applying some of the technology of war to civil uses even before the company came into formal existence. Underpinned by a £250 000 contract to make the military Pye 62 set for India, Woolgar was already developing two-way radios for both the police and new commercial users.

Telecomm was John Stanley's entry into the world his father had created, an entry both men seem to have thought inevitable. Though often absent as a father, C.O. had taken care picking boarding schools for his son. After Stowe, which he liked, John went to Cambridge to read Mechanical Sciences at King's College, and did war service in the Navy as a noncommissioned radar instructor. Finally C.O. sent him to work for a year in America with the telephone manufacturer Stromberg Carlson. John's mother, whom he loved, had died in 1943, leaving him to Velma's deft, if demanding, supervision.

He was 23 when he became a director of Telecomm, and moved into the Pye guest house where a housekeeper and butler looked after him. He had a car (still a rarity in those days of austerity) and impressed his cousins with champagne and oyster parties. He did not look like his father. Gangling and short-sighted, he was clumsy with people he did not know and lacked C.O.'s easy ability to charm. Peter Threlfall thought John 'shy, introvert and trusting . . . in many ways the antithesis of his father'. B.J. Edwards took against the young man, whom he called 'the weed'. Even Eddie was heard to say his nephew needed 'a kick up the backside', though he later changed his mind and became John's unqualified supporter, which could not be said of C.O. Word went round the family and among those who worked closely with C.O. that he found his son a disappointment. C.O.'s private secretary Daphne Whitmore thought him so 'blinkered' about founding a dynasty that he expected a son to be a 'replica of himself', and was often baffled by the real young man before him.

Had C.O. been able to look at his son more dispassionately, he might have seen that Telecomm was made for a young man who had an aptitude for mathematics, a good scientific grounding and also, it seemed, a capacity for original thought. Certainly he overwhelmed even Donald Jackson with his passion for the new technology of communications. 'He was interested in the technical side of the business in a way that C.O. never was, . . . [and on radios and telephones] you couldn't stop him. He talked about their future in cars, in your pocket. We thought him a bit mad.' After two months at Telecomm John saw there had to be a division of responsibilities between himself and Woolgar. John's version of this tricky moment was, 'I said to him, honest to God, Harry, either you must run the factory and I'll run the sales or you run the sales and I'll run the factory'. The upshot was that Woolgar became Telecomm's salesman and John took over development. By the end of 1948 he was chairing meetings of the Telecomm board in Eddie's absence.

The war left Pye well-placed to exploit an untapped market in radio telephones. Up to 1939 the only mobile radios in Britain belonged to the police, but the system was extended during the war to 300 fixed stations and 2000 mobiles giving both police and fire services coverage of the country. Pye had its first success a year before the formation of Telecomm when the Home Office granted the Cambridge taxi firm Camtax Mobile Licence No. 1. Equipped with Pye's PTC 202 radio the Camtax fleet were Britain's first radio-equipped commercial vehicles. The radio, a development of Pye's famous tank set, was mounted in the car boot. It gave good reception but used so much power that a careless driver could drain his battery if he transmitted for any length of time.

John later thought of these first commercial radio telephones as 'bloody great boxes and . . . not very good', but when Telecomm tried to improve on them it designed so much sophistication into details such as plugs and sockets that they cost more than the rest of the equipment together. The breakthrough came with the Reporter, a set that reduced the two-way radio almost to a single unit, and remained in production for over 15 years. This allowed Telecomm to install the first radio

Figure 6.19   Early Pye radio-telephone equipment, 1947

in a racing car (for Peter Clark of the HRG team in the 1949 season at Silverstone), and later in Prince Philip's Jaguar, making it the first radio-equipped royal car (Telecomm engineers inadvertently drilled a hole in the car's fuel tank and discovered the mistake only when it drove off leaving a trail of petrol). C.O. had passed on to his son his love of the sea, and John's passion for sailing gave him the chance to test each new Telecomm ship-to-shore radio in his own yacht.

C.O. was himself involved in the best publicity coup, the equipping of Sir John Hunt's 1953 expedition to Everest. George Band, the Cambridge undergraduate responsible for the team's communications, approached Pye for help and C.O. at once offered full support. Hunt

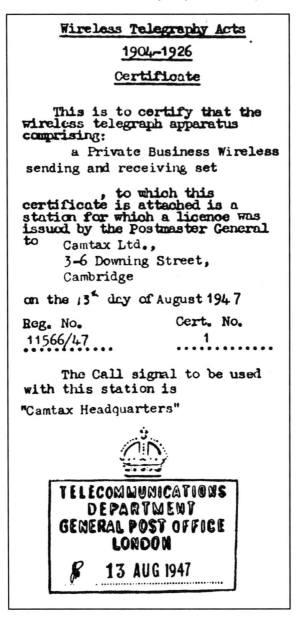

Figure 6.20

took two of Telecomm's short-wave receivers and eight walkie-talkies, all adapted to operate in extreme cold. The climbers used the receivers to listen to weather reports and, for relaxation, bizarre commercials from a radio station in Ceylon. The walkie-talkies provided communication

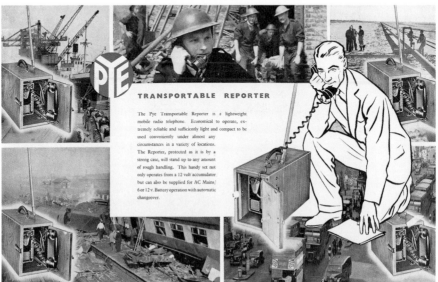

Figure 6.21   Leaflet for Pye's successful mobile radio telephone, 1954

between camps. The sets worked well up to 24 000 ft but the weight of the batteries ruled out use on the final ascent. The result was a publicity coup for Pye as well as the British team. News of the ascent coincided with the Coronation, allowing the *Daily Express* to shout on its front page, 'ALL THIS AND EVEREST TOO'. C.O. loved it.

One of Telecomm's first and biggest jobs was the development of ILS, the Instrument Landing System that made it possible for aeroplanes to land in poor visibility. B.J. Edwards had discussed the feasibility of blind landing with Robert Renwick's 60 Group in the war, and in 1944 the government Telecommunications Research Establishment passed all work on the project to Pye for development. ILS used passive transmitter aerials that radiated pulses, allowing cockpit instruments to show if the plane was on course and at the right angle of descent. In 1948 John had the unnerving honour of flying in the RAF Oxford that made the first test flight of the system. 'We did land with no hands, no question about it, but the last fifty feet downwards were, to say the least, terrifying.'

Originally ordered for the RAF and British airlines, the requirements for ILS became increasingly complex as it was sold around the world and came under international regulation. But it was a rich market in which Pye was leader. By 1950 work on ILS tied up a large proportion

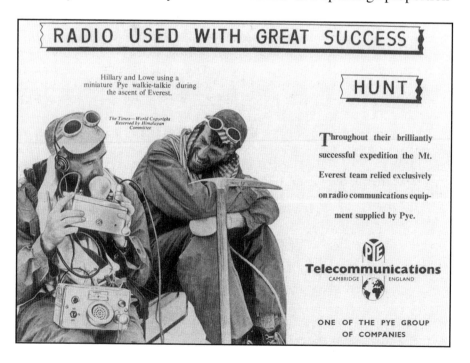

Figure 6.22   Pye share in the celebrations on the conquest of Everest, 1953

Figure 6.23   Engineer Geoffrey Peel with an early design of the Pye Instrument Landing System antenna at RAF Martlesham Heath, c.1946

of Telecomm's resources, but it also contributed to Telecomm profits which the following year reached £100 000 for the first time. In 1953 Telecomm leased a new factory at the Ditton Works built by Pye Ltd. at a cost of £70 000. By 1955 it was the dominant supplier of mobile radios worldwide with over 80 per cent of all police forces outside the United States using Pye equipment (the Americans operated on FM, Pye's equipment on AM).

John said of his first years at Telecomm, 'I virtually had a little empire of my own. I was just given the job of building up telecommunications and told to get on with it. For 5 or 6 years my business contacts with [my father] were relatively small'. Others did not see it quite like that. 'It was difficult for [him]', Donald Jackson thought, 'because right to the end he may have been supposed to be in charge [of Telecomm], but we all knew that C.O. was still the man'. Sometimes the father displayed his power in the crudest way. Not long after putting the first mobile radios into British taxis, Telecomm won a contract to equip taxicabs in Paris and the directors took their wives for a week's stay in the French capital to celebrate the inauguration of the new service. According to John's secretary Marjorie McCarthy it 'offended' him that directors always got these perks, and he had eight lower-level Telecomm employees flown out to enjoy the occasion too. When C.O. heard that what he called 'nonessential staff' had been on a foreign 'jamboree', he ordered John to pay back the cost from his own pocket. It was perverse, for C.O. was

Figure 6.24   Front cover of a leaflet, 1959

inclined to such gestures himself, and humiliating for John as word of what happened went around the factory.

In one matter father and son were passionate, and almost equal, partners: running Telecomm turned John into as determined an enemy of

Figure 6.25   Mobile radio communications equipment being manufactured at the vast Pye Telecommunications factory, Cambridge, c.1955

government controls as C.O. Telecomm's future depended on expanding the market for two-way radio, yet every area it wanted to move
into – the ambulance service, ship-to-shore communications, the railways – required the Post Office to grant the licensed use of a frequency.
This the Post Office was often reluctant to do, for, armed with the
Victorian Telegraph and Wireless Telegraphy Acts, it guarded the
nation's airwaves with the vigilance of a gamekeeper on a ducal estate.

The pattern of battle was set before the official launch of Telecomm
when Harry Woolgar sought a two-way radio licence for a firm operating
tugs on the river Tyne at Newcastle. The Post Office was so obstructive
that C.O. threatened to have the matter raised in the House of Commons
and the licence came through only within minutes of the deadline he had
set. John Stanley came to the conclusion that the Post Office thought
'everything to do with communications was their responsibility and
everybody else . . . was stepping on their hallowed ground'. This had the
advantage for Pye of frightening off the weak-hearted among potential
competitors, while it drove C.O. into a sustained campaign of Post
Office baiting.

In Pye's annual reports the Post Office began to play the part of
a pantomime villain. An article with the title 'Grandfather's Progressive Outlook' pointed out that when Parliament passed the Telegraph
Act in 1869 telephones had not been invented. It described how when
Telecomm first offered two-way radios to ambulances in 1949 the Post
Office insisted there were no wavelengths available. Months of nego-
tiation produced first one, then three, but it was not until 1951 that
'we were given reason to believe that sufficient wavelengths would be
available to serve all the boroughs and counties in the country'. And
someone paid a high price for the delay. An injured child had died
while being taken round Bradford by an ambulance looking for a suitable hospital. The coroner recommended that all ambulances be fitted
with two-way radios.

A memorandum written by a senior Post Office civil servant shortly
before Labour lost power in 1951 suggests the mix of rage and confusion
C.O.'s taunts caused in Whitehall. The writer insisted there was 'little if
any factual basis' in what C.O. said, and many 'factual distortions'. At
a recent meeting with the Post Office C.O. had made 'numerous allegations', and then orchestrated half a dozen questions in the House of
Commons all demanding information 'of no outstanding public interest'. The motive was political, the memorandum concluded, and since
C.O. distorted any information that was given him, and criticised 'for
criticism's sake', his request for a meeting with the Postmaster General
should be refused. On that, at least, all at the Post Office were agreed.
One minute attached to the memorandum argued that a meeting would
only inflate C.O.'s 'self-importance'. Another observed that 'nothing
will restrain Mr Stanley especially when he gets on his feet to make

Figure 6.26

a speech'. This writer admitted he was puzzled. C.O. might be 'very much a lone hand', but he was also chairman of a 'progressive and competitive company'. The best thing was to leave officials to 'battle [with him] as best they can'.

Both sides battled, but in 1954 Pye battled best, though the victory was as much a surprise to C.O. as it was to the Postmaster General. By this time C.O. had made more enemies than usual by his campaign for commercial television described in the following chapter. When the Post Office solved the problem of finding frequencies for the new TV channel by taking them from two-way radio, C.O. saw it as an attempt to punish Pye, and he instructed his lawyer Frank Levinson to search the Wireless Telegraphy Act for evidence that the Post Office was acting outside its terms. Levinson drew a blank there, but reported something very strange. As far as he could make out the Act had never been laid before the House of Commons. If that were so, the Post Office had no authority to issue licences of any kind.

For once C.O. acted with discretion. Pye persuaded a Colchester firm that was a customer of Telecomm to start proceedings against the Post Office to recover the money it had paid in licence fees. Telecomm agreed to pay the legal costs and hired the well-known barrister Sir Hartley Shawcross. As soon as the case came to the Old Bailey the Post Office pleaded guilty, and agreed to issue alternative frequencies to users of mobile radios. What C.O. did not understand until later was that the Post Office's greatest anxiety was not two-way radio, but the millions of licences it had issued over the years, apparently illegally, for ordinary radio and television sets. (The government moved quickly to correct this with retrospective legislation.)

Whitehall was not the only source of C.O.'s frustrations. A large part of the Pye report for 1954–1955 was taken up by an attack on British Rail for its failure to appreciate the usefulness of mobile radio. The fashionable young cartoonist Marc Boxer provided a drawing of an engine driver releasing a pigeon with a letter in its beak, and an article with the title 'Puff-Puffs' summed up C.O.'s scorn for the men who ran Britain's railways. France, he pointed out, had equipped its newly invented fast train with radios. One hundred and thirty railroads in the United States used radio communication over 47 000 miles of track. What had British Rail done? It had conducted a 6-year test of a radio network in half of its biggest marshalling yard, but would not extend it to the other half. Pye had proposed installing trial equipment on a stretch of passenger track. 'Monotonously [British Rail's] verdict is that the risk is too great because the presence of a radio-telephone might distract the driver.'

At times it seemed C.O. could not lead Pye anywhere without getting in a scrap. At the end of the war the BBC and a large part of the radio industry expected broadcasting to move out of medium and long wave into very high frequency (VHF) to take pressure off the increasingly crowded airwaves. VHF broadcasting offered a choice between two systems of modulation, amplitude (AM) or frequency (FM). Most British experts favoured FM; C.O., advised by B.J. Edwards, insisted that AM was the sensible way forward. In 1952 he managed to be appointed as one of two industry representatives on the Television Advisory Committee (TAC), which was charged by the government with deciding, among other matters, between AM and FM.

B.J. Edwards, who was a member of TAC's technical subcommittee, thought Pye's technical arguments strong enough to win over the Committee's chairman Admiral Sir Charles Daniel and one or two other members, and judged that C.O.'s campaigning had 'caused sufficient confusion in the BBC and the Post Office to give the government the opportunity to squash' the FM lobby. One of C.O.'s contributions to the confusion was an account of Australia's problems with FM sent him by his Australian business partner Arthur Warner. Australia had set up two FM stations in 1950 when, in Warner's words, there was 'a wave of theoretical enthusiasm' for FM and 'the politicians, who of course understood as little as usual on such subjects, were overwhelmed with the "bally-hoo".' But few people listened to the stations any more, evidence enough for C.O. that FM was, or ought to be, a dead duck.

His salesman's instinct told him FM would not be popular. Most radio owners showed little interest in it, and certainly would not want to pay an extra £10 to buy the more expensive FM sets if the only advantage was somewhat better reception of existing BBC programmes. (Poor sales of the new sets after the BBC began FM broadcasting suggested that C.O. was right.) He also feared an FM broadcasting system would compete for funds at a time when 'every penny available should be solidly behind the expansion of television'. The latter was his obsession and so, increasingly, was the development of local, and he hoped commercially funded, radio which he feared would also suffer if the BBC's national broadcasting went over to FM.

He failed to convince his colleagues on the Television Advisory Committee, which plumped for FM , though it did call the introduction of VHF broadcasting 'an unwelcome complication'. C.O. could only repeat his counterarguments in a long minority report, but his noisy opposition helped delay the introduction of VHF broadcasting until 1955.

There were other battles in which C.O.'s pride was as much engaged as Pye's commercial interest. One might have expected the company's wartime achievements to have ensured its future as a defence contractor, particularly when fear of the Soviet Union and the onset of the Korean War drove the British government to re-arm. According to a Ministry of Supply survey of defence-related industry, post-war Pye and its associated companies had a development team of over 200 engineers and scientists. Important contracts were obtained, including radio for the Navy and communications systems for NATO, but C.O.'s insistence that he knew best soon led to wearisome altercations familiar to anyone who remembered World War II.

Due to Purchase Tax increases, the following
prices now apply:

FenMan I   -   30 gns. Tax paid
FenMan I RG -   68 gns. ,,   ,,
FenMan II   -   40 gns. ,,   ,,
FenMan II RG -   89 gns. ,,   ,,

FM/AM RADIO

Figure 6.27   Front cover of a leaflet, 1955

Often he seemed to delight in a fight for fighting's sake, as in the ill-tempered war he fought with Whitehall over the C40, the new radio for armoured fighting vehicles chosen to replace the Pye 19 set. Pye accepted this contract in full knowledge of the specifications laid down by the War Office, but at the end of 1950 C.O. told the Ministry of Supply the new set was too complex, and too expensive, to make. He recommended scrapping it, and suggested Pye adapt the wartime Pye 62 set to take its place. The Minister of Supply, Geoffrey Strauss, entered the dispute and asked why Pye had not raised the alarm earlier. C.O. offered the disingenuous excuse that he had not thought it proper for 'a commercial organisation to press its opinion'. Ministry officials were furious – 'Mr Stanley does not know what he is talking about' – and wondered if C.O. was losing interest in the contract because of his growing involvement in television and civil radio communications (a reasonable suspicion, though C.O. provided no evidence for it).

But they also knew they depended on Pye for the C40 and that it was 'unprofitable to enter polemics' with C.O. Stanley. Strauss promised him whatever help he needed to complete the contract.

A short while later the Ministry of Supply offered Charles Harmer the newly created post of Director-General of Radio Production with a brief to co-ordinate government contracts for radio, radar and telecommunications equipment. Harmer may have been tempted to accept, but C.O. reacted as though it was a trick to neutralise the ever troublesome Pye. One may also guess he could not stomach losing an old colleague to the 'little men' in Whitehall. It was Harmer's decision, C.O. told the Ministry, but pointed out it would be difficult for him to take up the new job until its dispute with Pye was settled. He also told Harmer the choice was his, but in a way that made it hard for him to say yes. Harmer should remember that 'in some quarters you are considered too 1939 war minded'. Could he be sure he would have any influence on the design of new equipment? Mightn't the job amount to no more than 'working out' the mistakes of others?

Harmer stayed, and C.O. continued his battle over the C40 with the new Tory government that came to power in the autumn of 1951. When he could not get to see the new Minister of Supply, Duncan Sandys, he lost his temper and complained to the Conservative party chairman Lord Woolton. 'During the war... I once arranged for twenty nine questions to be asked in a week... with the result that eventually the Prime Minister ordered the Du Parcq investigation... I had hoped to avoid having questions [on Pye's current problems asked] in the House.'

C.O.'s version of that episode was questionable, but the threat worked. Woolton knew C.O. well enough as an active Conservative supporter to want to please him, and recommended him to Sandys – 'I think highly of [Stanley] and personally I would value his advice'. Sandys received him and agreed Pye should be allowed to withdraw from the C40 contract. But C.O. still plagued the Minister, agreeing to serve on the new Radio Re-Armament Advisory Committee and then telling Sandys he would 'certainly come adrift' if he relied on its members to prevent Britain acquiring 'an electronic Maginot Line'. Ten years later John Stanley explained that Pye had taken few post-war military contracts because '[we] won't make rubbish'.

C.O.'s determination to do what he wanted regardless of others was the reason for Pye's private programme to make its own anti-tank missile. He had always liked the idea of Pye developing military equipment on its own, free of the restraints of a government contract, as it had briefly done with the proximity fuze during the war. Pye's wire-guided anti-tank missile was a development of the plywood rocket that B.J. Edwards experimented with at the start of the war before passing the designs to America (Edwards kept copies, and these were the basis of the new weapon). C.O. argued that a freelance venture was the only way for Pye to break into missiles, but when he revealed his project to the Ministry of Supply in 1953 the response was unenthusiastic. Officials doubted Pye had sufficient knowledge of aerodynamics, or realised the likely cost. They applauded C.O.'s enthusiasm but did not want Pye to get into 'difficulties which can be avoided by careful planning'. Undeterred, C.O. obtained test facilities at the Royal Aeronautical Establishment in Farnborough, where the weapon was test fired and later displayed at the Air Show. The press release described it as 5 ft long, weighing only 80 lbs and with a range of 1 mile, and Pye announced it was ready to go into production as soon as orders were received. There were none, and when C.O. was asked about the project at a board meeting in 1962 he said he was spending no more money on it.

Ten years of rapid post-war growth seemed to give a final seal of approval for the way C.O. ran the company. The low capitalisation brought out so clearly by the government's 1949 survey of the industry reflected two of his guiding principles. Pye's success depended on C.O. Stanley remaining in unchallenged command; and the best way to get money was to charm or bully it from banks ('I am a believer in overdrafts', he remarked to a friend, 'as long as you can get away with it'). By 1955 authorised capital had ballooned to £5.2 million, growing by £3.2 million in the two last years alone, and C.O. was using shares to acquire subsidiaries, but the Stanleys still owned the majority shareholding. This was possible because Pye new issues, handled by Greenwells, the stockbrokers with which Robert Renwick was associated, usually took the form of A Deferreds that carried no voting rights. A Deferreds were typically used to raise £530 000 to buy a 52 per cent interest in Tecnico, the Australian electronics company owned by the politician turned businessman Arthur Warner, whose forcefulness C.O. much admired.

C.O. offered a defence of his strategy for raising capital to Pye's 1955 annual general meeting:

> Every director on your board has grown up with and worked in the company most of their working lives, in fact they are all company men. When the real expansion in your company began they realised that it was essential to prevent control of the company passing into the hands of some member of an international cartel ... or any group that might upset the continuity of management ... They were [also] determined to remain a British company.

If one read 'C.O. Stanley' instead of 'the directors' it was an accurate statement of the facts. But opinion in the City was turning against nonvoting shares and this strengthened C.O.'s interest in banks, for experience taught he could usually make money by borrowing money.

He did not want partners, so why acquire them even in the diluted form of shareholders with voting rights if he could treat a bank like a servant?

A brush with Barclays in 1948 suggests his technique. He asked its general manager, his old acquaintance G.F. Lewis, to raise Pye's overdraft limit to £900 000 to cover both Pye and its subsidiaries. Lewis infuriated C.O. by spending half their meeting explaining that such a large overdraft would be an 'unnecessary embarrassment', but agreed to allow the Pye group an overdraft of £1 million 'upon proof that steps were being taken to issue further shares'. C.O. took offence at this condition, arguing that a new share issue would mean offering 'the property which belong[s] to my shareholders [at] considerably less than its real value'. He threatened to publish his correspondence with Lewis and take Pye's business elsewhere, and made a show of talking to the Westminster Bank.

Barclays did not take fright; even its manager in Cambridge dared to question C.O.'s claim that the price of any new shares would not represent Pye's 'real value'. Wasn't value determined by the market? The bank's patience still held when C.O. challenged its version of what had been said at their meeting, and brought in Fred Keys to back him up. In the end he got a £1.5 million overdraft for the group but Barclays, convinced Pye needed more money to fund expansion than C.O. admitted, insisted on a quid pro quo. It was aware, it informed him, of 'your unwillingness that your companies should give security, and also your dislike of having guarantees shown on your companies' balance sheets'. Nevertheless to obtain the new overdraft Pye would have to make a debenture issue, maintain 'liquid surpluses' and agree to a share issue within two years. By 1955 C.O. had talked up Pye's overdraft limit to £3 million and was still, as he might have said, getting away with it.

He also got away with his style of management. At the end of the war he had made a broad division of responsibilities among directors. Pearl would advise on matters where 'owing to lack of information, the management might make mistakes in relation to the work-people'. Eddie's field was sales and service policy, Harmer's 'internal engineering and works problems'. C.O. had become chairman after the death of Sir Thomas Polson in 1946, and the death of Millward Ellis the same year left no one on the board apart from Pearl who was likely to challenge him. There was a curious incident at the start of 1947 when he insisted Keys alter the minutes of a board meeting to record a 'disagreement' between himself and Pearl. It was scarcely a grave matter – there was doubt whether she had told him that Invicta had released some television sets onto the market – and the fuss C.O. made over it suggested that for some reason he wanted to strike a pose of correctness.

The following year something more serious happened. Doubts seem to have been raised, either in board meetings or out of them, about C.O.'s domination of the company. C.O. told the four surviving directors

(Pearl, Eddie, Charles Harmer and L.G. Hawkins) that he proposed establishing a group board made up of a director from each of the growing number of subsidiary and associated companies. This would meet under his chairmanship 'a few times every year'. He would also set up a Pye executive board to 'build up those executives . . . responsible for the running of different sections' while 'improv[ing] their status' and preparing them for eventual membership of the main board. The minute of the meeting continued:

> Mr Stanley referred to the statement which had been made to him from time to time that the company was too much of a "one-man business" . . . Mr Stanley also stated that he knew of a man who could come into the business and completely take over from him in a very short time if the directors preferred this as an alternative to the executive board. The directors expressed themselves entirely opposed to any such suggestion as they were very happy about the present arrangement. They also expressed doubt that the man Mr Stanley had in mind, or anyone else, could adequately fill the managing director's position which was largely built up around Mr Stanley's personality.

Did C.O. mean it? Did any of those listening to him believe he meant it? Keys, who as company secretary kept the minutes, seems to have been not quite sure. In his handwritten draft account of the meeting, after noting C.O.'s offer to stand down, he added, 'No Go!' – certainly a reassurance to himself, if not a full-blooded cry of alarm.

John Stanley, B.J. Edwards and Norman Twemlow joined the main board in 1951, but while they added energy they were no counterweight to C.O. Pye's achievements depended less on its board than on the relationship between C.O. and the men who ran the company's different parts. When John Gorst (later a Conservative MP, and knighted in 1994) joined Pye as press officer in 1953 his new employer never told him what to do, just asked what he had achieved. Gorst found C.O. stimulating and always 'positive'; if he said no to an idea he invariably offered an alternative. 'None of us thought Pye good payers but we accepted that because it gave us responsibility and opportunity.'

Gorst decided C.O. liked to employ people who could 'pick themselves up after a setback'. C.O.'s talent, another executive thought, was for 'picking a man, giving him a task, and giving him a little bag of silver and saying "go out and double it." [It was] almost biblical.' This was how C.O. had – for much of the time – allowed John to develop Telecomm. The worst recommendation for a job at Pye was to have worked for the government. C.O. turned down a senior officer from the National Fire Service who applied to be transport manager on the grounds that it was 'much too risky to take anyone who has had ten years of form-filling experience'.

He seldom used money to motivate people. Both he and Pearl expected executives to live modestly, and were known to inspect the factory car park to check that employees' cars were not too grand. C.O.'s recipe, apart from giving people their head, was charm, inspiration and the occasional display of temper, and many of those who worked for him thought it brought results. He received 'the most extraordinary loyalty because he trusted and inspired people to do things they never would have imagined they could do' (John Chilvers, who joined Pye as an apprentice). 'He was prepared to trust the young and give them a go. People loved him for that' (Norman Leeks, another apprentice who stayed at Pye all his working life). Staff might be in awe of him, but not because he put on a grand manner. He remained approachable; and still turned up at the factory in shabby clothes. It was remembered with affection that he went to a Pye Sports Club dance with paper clips doing duty for cufflinks. When there was a crisis and people had to work throughout a weekend they did it for no extra pay because C.O. would be there too.

John Stanley's secretary Marjorie McCarthy thought C.O. still ran the company 'as a family'. If so, it was a family so unstructured that it looked chaotic, and that was what he wanted. He explained at the end of the war that he was determined to avoid a 'cast-iron' organisation that would 'hinder the progress and initiative of individuals'. Pye had no organisational charts. Office doors bore neither names nor titles. Internal communication was informal, and more by conversation than by memoranda. Order came from C.O. as the family's patriarch, though as the years went by an increasingly absent one.

In the early 1950s he would spend week-ends at Sainsfoins, then work in Cambridge on Monday, staying for the executive board meeting the following morning. Then he went to London for the rest of the week where he was occupied with manoeuvres over the future of British television and other matters that interested him more than the routines of factory production. His youngest sister Rue said he could walk into an office or factory and know at once if something was wrong. Perhaps he thought Cambridge needed no more attention than that. He was also travelling abroad more often with Velma, developing a pattern that took him almost every year to Australia and New Zealand for business and sometimes deep-sea fishing, and then for a holiday to Jamaica. The itinerary might also include America, which he liked to scan for auguries of Britain's future. These winter journeys could last three months, and another month was spent in Ireland each summer. Given the poor state of long-distance telephones it meant he was out of easy communication with Cambridge for at least a quarter of the year, leaving day-to-day management to the executive board of six. The main board, reinforced by executive directors, was supposed to take care of strategy, but it met irregularly when C.O.

Figure 6.28   *C.O. Stanley's success at deep sea fishing, New Zealand, 1955. His son John, with his wife Elizabeth, shared the same enthusiasm*

was away, and anyhow was unlikely to take important decisions in his absence.

The problem of establishing financial control over an expanding business was compounded by C.O.'s preference for setting up new ventures as subsidiaries rather than as divisions within Pye, and also by his failure to appoint a powerful accountant to the company's staff. He preferred to make do with the amenable Fred Keys, who knew C.O. 'disliked finance directors and accountants'. Sometimes C.O. looked about to change his mind. In 1948 he told the board that, subject to Keys' approval, he proposed appointing a Scots accountant, 'a man of excellent report, . . . [to] control the figures of the whole group'. The Scot 'of excellent report' was Gordon Maclagan, the 'Renwick boy' whom C.O. had worked with and admired during the war. C.O. despatched the odd couple of Keys and the salesman Norman Twemlow to offer Maclagan the position of Pye's chief accountant. This they did, but at the same time warned him not to tell anyone about it. Thinking the secrecy odd, the Scot turned down the offer.

Perhaps C.O. was never serious about it; perhaps Keys wanted to deter this potential rival, for he was famously jealous of his closeness to C.O. Whatever the reason Pye remained without a financial controller and hands were wrung at board meetings about overheads that were beyond control. In November 1951, after C.O. had left on his winter travels, Pearl instructed the main board to cut overheads by 25 per cent to compensate for poor trading conditions and tighter limits on hire purchase. Four months later directors received an analysis of the previous 4 years showing a steady rise in overheads and a substantial drop in the rate of gross and net profit. L.G. Hawkins thought conditions bad enough for the board to meet every fortnight, but nothing was done either about that or his suggestion that savings could be made by bringing subsidiary companies under one administrative roof, and centralising office and sales activities (this was rejected on the grounds that the 'necessary premises' were not available). At the end of 1955 C.O. declared the drive against overheads a failure and said he would have to appoint one person to work full time on the problem.

Overheads seemed a minor matter in years that were generous with profits and prizes. In November 1953 C.O. gave a grand dinner at the Dorchester hotel in London to celebrate the government's award of £20 000 as settlement of Pye's claims for original work done in the war, but never properly rewarded. After lengthy representations by Pye the Ministry of Supply agreed to single out two projects – the valve for the radio-operated fuze, Pye's precursor of the proximity fuze, and the No. 19 set used in tanks of all the allied armies (C.O. split the money between the team that developed the No. 19 set led by Charles Harmer, and B.J. Edwards' fuze team). In his speech that night C.O. talked about his company's achievements to a large audience of politicians, senior officials past and present and leading scientists, among them Sir John Cockroft. He was in fluent form, and took delight in making his favourite point that under cover of the war America had stolen credit due Britain in general and Pye in particular. There was, though, no nostalgia in what he said. War had propelled him and Pye to prominence, but peace seemed to promise greater triumphs still.

Sources

COS files: 1/8/1–2, 2/15, 2/17 (Pye annual reports), 2/2, 2/4–5 (quarrels with Ministry of Supply) 2/5 and 2/7 (anti-tank missile project), 6/1 (B.J. Edwards' 1944 paper on post-war TV), 7/1/6.

Simoco boxes: Box 1 (board meeting agendas 1940–1946, including Stanley's 1946 remarks on difficulties of transition to peace; Fred Keys' copy of the main board minutes 1947–1955 with the most detailed version of Stanley's 1948 offer to stand down), Box 2, Box 4 (Institute of Directors 1952–1956).

Pye main board minutes, Pye Telecomm minutes 1948–1955.

Public Records Office: HO 255/189 (Pye's 1951 attack on GPO), HO 256/02 (survey of TV manufacturers), HO 256/208 (Stanley urges TAC to adopt 625 lines).

Interviews: Brian Beer; Michael Bell, James Bennett, Sir John Clark, Mike Cosgrove, Richard Ellis, Jo Fletcher, Dennis Fuller, Sir John Gorst, Donald Jackson, Fred Keys, Jim Langford, Marjorie McCarthy, Geoff Peel, Ian Sichel, Peter Threlfall, Willie Wakefield.

Interviews: Michael Bell; Alan Bednall, Sam Carn (Cripps' visit to Cambridge), Les Germany, John Stanley (Telecomm), Harry Woolgar (Telecomm), Michael Worsley.

A–Z files: Lord Renwick.

Books: Barnett, *The Lost Victory*; Bell, *ibid.*; Briggs, *ibid.*, Vol. IV; Bussey, *The Story of Pye Wireless*; Cairncross, *Years of Recovery*; Dow, *The Management of the British Economy 1945–60*; Geddes and Bussey, *ibid.*

Chapter 7

Liberating television

Robert Renwick, who never believed that work had to be done in an office, conducted much of his business at the Savoy Grill, where he ate lunch almost every day. Once a week when C.O. was in London he joined Renwick there, and over bottles of their favourite Pouilly-Fuissé the two men plotted the future. It was one of these lunches, in 1949, that sparked the idea of a campaign to break the BBC's monopoly of broadcasting.

They were talking about a subject that obsessed C.O. – the slow sales of television sets. The war had scarcely ended when C.O. began thumping the drum for a rapid expansion of British television. Two months before the BBC resumed transmission in 1946 he lectured the British Radio and Electronic Equipment Manufacturers' Association (BREMA)'s television promotion committee, which he chaired, on the need for Britain to win back the world leadership in television it had enjoyed before the war: other countries 'may have dollars, they may have Communism, but we want television'. It was a 'new art', a valuable export, and a creator of employment. It was also the child of private enterprise, which was not to be forgotten at a time when a Labour minister (Emmanuel Shinwell) could say that the British people 'were in no mood to stand any nonsense from private enterprise [and] if industry cannot deliver the goods the government will do so'.

C.O. promised that as far as television was concerned the industry would 'in spite of Mr Shinwell's nonsense . . . deliver the goods', but at the lunch with Renwick three years later he could only ponder on gloomy statistics. In America there was nothing unusual about a house with a TV aerial. Television was everywhere and, as the British journalist Alistair Cooke reported, already 'as humble as a hot dog'. In Britain it was still caviar. The BBC, partly because of government-imposed austerity, had not yet built transmitters outside London, and even after the Sutton Coldfield transmitter brought television to the Midlands in

Figure 7.1   Front cover of a magazine for Pye dealers, first issued in 1946. Note that the right-hand segment shows a Pye model B16T television receiver on a shop counter

Figure 7.2   Queen Mary with C.O. Stanley at Radiolympia, 1947

December 1949 the spread of TV sets remained slow. Some people suspected this was what the BBC wanted. The Corporation's main interest remained sound broadcasting and, at the time of C.O.'s lunch with Renwick, it was forecasting there would be only 2 million television licence holders by 1955 (the actual number would be more than double).

Pye News

A BULLETIN OF INFORMATION FOR ALL PYE WORKERS

No. 12   Vol. 1   12th November, 1948

PYE ON THE BEAM

"Coming in on the beam" is an American expression much used in films about those fearless aviators who, we are led to believe, practically won the war on their own. The beam is, of course, a radio beam, and, like everything else connected with radio, of interest to Pye. We call it an Instrument Landing System (I.L.S.), and here at Cambridge work is proceeding in accordance with a plan which, when it is completed, will mean that a pilot will be able to land his machine on any major aerodrome in any sort of weather.

I.L.S. are still rather the back-room boys, and if you were to ask the average Pye worker what they do—well do you know ?

What I.L.S. Does

I.L.S. is a system of transmitters, five in all, which send out two beams. One beam shows the pilot if he goes to the left or to the right of the line on which he has to fly to land properly. The second beam, which may be likened to a searchlight shining up at an angle, tells him if he is coming in too steeply or too "flat." All the pilot

Dickie Attenborough at Pye

STRICTLY **NO ENTRY** FOR **ANYONE** OTHER THAN PYE EMPLOYEES

We were glad to be able to print a picture of Mr. Richard Attenborough in our previous issue, but we are sure all PYE NEWS readers will be happy to see these snapshots of the famous actor at the factory. "Dickie's" show "Home of the Brave" at the New Theatre was a great success, and many Pye people enjoyed the show during its week's run. We hear that it is due to open in the West End shortly, and some of those departments who want chara. trips to Town and a show might remember the name when they are making their plans.

Figure 7.3   Richard Attenborough with a Pye B18T television receiver at the Pye factory, Cambridge, 1948

The slow TV sales were particularly frustrating for C.O. because Pye's superior sets were likely to ensure him the lion's share of a bigger market. He was already planning a similar lead in colour television, but to finance this needed greater profits from sales of black and white sets.

C.O. argued that TV sales would only go up when the BBC showed more interesting programmes. Earlier that year he had made one of his trips to America, and he described to Renwick the variety of programmes produced by a broadcasting system based on competition and financed by advertisers. Renwick, scarred by the nationalisation of his family business, was an easy convert to the idea of ending the BBC monopoly and by the end of the lunch they had resolved to campaign for what they called 'competitive television'. C.O. would provide the money; Renwick the contacts in the City, the Conservative party and the newspapers whose own dependence on advertising revenues made them likely opponents of advertising on the airwaves.

The debate over whether part of British television might be financed by advertising came to seem like a battle for the soul of the British people: one of the combatants thought it the greatest national controversy since the revision of the Church of England Prayer Book in the 1920s. C.O. may not have foreseen the intensity of the coming storm because he knew that introducing a commercial element into British broadcasting was not a new idea. The expense of pre-war television (it cost the BBC £450 000 a year to service some 20 000 licence holders) led the Postmaster General's Television Advisory Committee (TAC) to report in 1939 that 'in view of the great difficulty in financing the [BBC] service ... we consider that the inclusion of sponsored programmes and even direct advertising ... would be fully justified'. Both the film industry and the radio manufacturers agreed, and perhaps only the war prevented a government move in that direction. When C.O. led a Radio Industry Council delegation to give evidence to the Hankey committee towards the end of the war no one was shocked when he argued for some sort of commercial broadcasting. Hankey's 1945 report judged that business would not be interested in television advertising until there was a bigger audience, but stressed it wanted to make this point 'without prejudicing the matter for the future'.

The climate had changed by 1949, when the Labour government charged another committee, this time under Lord Beveridge, to reconsider the future of television. Opinion polls suggested that only a third of the public were interested in commercial broadcasting, while a large part of the country's opinion-makers hated the idea. The war had enhanced the BBC's reputation, and those who saw themselves as guardians of the nation's spiritual well-being – churchmen and educators as well as many politicians – disliked what they knew of post-war American television. There was also apprehension, from which C.O. himself was not immune, about the effect of an American-dominated world on the British way of life. Even part of the electronics industry opposed changes in broadcasting, and when C.O. took a Radio Industry Council team to give evidence to Beveridge, several leading companies, among them Ekco, EMI and STC, distanced themselves from him. '[C.O.] was there first', Sir John Clark, chairman of Plessey, conceded later. 'He was right before the rest of us woke up to it.'

Supporters of commercial television did not agree on the form it should take. C.O. made the case to Beveridge for splitting sound and vision into two separate corporations in order to rescue television from the domination of sound radio within the BBC. He estimated it would cost £9 million a year to make good television programmes and that only sponsorship could raise such amounts. Sponsors would also be able to bring sporting events like the Derby to television, and viewers would get 'what they really did want and could appreciate' rather than what it 'was thought they wanted'. Appropriate moral standards could be maintained by broadcasting sponsored programmes within the framework of the BBC.

C.O. got another chance to influence the Beveridge committee when one of its members, the headmaster of Winchester, Walter Oakeshott, spent a day at Pye, but he spoilt it. He kept his important visitor waiting for half an hour before treating him to his standard lecture on the benefits of commercial (by which he meant sponsored) TV, and the need for programmes that were popular rather than high-minded. He then took the headmaster to a private room in the University Arms hotel where Pye's directors and senior executives tucked into a lunch prepared by a chef who seemed not to have heard of food rationing. There was gin and whisky, white and red wine, brandy and cigars. Oakeshott had a small glass of dry sherry.

After the most thorough enquiry into British broadcasting so far Beveridge upheld the BBC's monopoly. In its report published in January 1951, 7 of the 11 committee members, among them the headmaster of Winchester, ruled there was no case for ending the ban on advertising. Beveridge himself and two other members left open the possibility of a controlled advertising hour on the BBC and suggested that the public might be asked if it wanted to pay bigger licence fees to keep their television screens pure of advertisements, or raise revenue through advertisements and leave the licence fee unchanged. Only the Conservative MP Selwyn Lloyd argued in a minority report for a commercial rival to the BBC, emphasising the wider dangers of monopoly. 'The evil lies in . . . the control by a monopoly of this great medium of expression . . . the concentration of great power in the hands of a few men and women.'

The example of Soviet Communism and fears that an Orwellian Ministry of Truth might one day manipulate British minds were prompting people to re-examine the role of the BBC as a political monopoly. This was one of C.O.'s lines of attack when he delivered his verdict on Beveridge in Pye's 1951 annual report, condemning the committee members as 'proponents of twentieth century state worship' who showed 'sycophantic adherence to the totalitarian way of life'.

His words may have been inspired by one of the young Conservative parliamentary candidates C.O. was now employing, among them the markedly right-wing Anthony Fell, who became manager of Pye's London office 'to assist in the campaign for the more rapid expansion of the television system', and Kenneth Lewis, for a short time managing director of Arks. But C.O. needed no guidance to decide that the Beveridge report was biased against private enterprise and profit, or that it threatened to 'ruin broadcasting for the British people' by forcing on them 'a high standard of "culture" but a low standard of entertainment'.

C.O. did not let Beveridge interfere with his campaigning and in September 1950 brought a group of London advertising agents to Cambridge to show them what British commercial television might look like. After dinner at the University Arms the advertising agents turned to 20 television sets installed by Pye engineers to watch an hour-long variety show by some of Britain's most popular entertainers, among them Richard Murdoch, Fred Emery and Annette Mills with her puppet Muffin the Mule. The show was live, as most television then was, except for filmed excerpts of a Bruce Woodcock title fight which the BBC would not pay to televise. After the show C.O. took his guests to the factory to see Pye's own transmission tower and the studio equipped by Pye, and operated by a dozen of its technicians.

The demonstration was possible because B.J. Edwards had convinced C.O. that Pye would sell more television sets if it was involved in the whole television chain from transmission through cameras and studio equipment to the domestic set. The easily bored Edwards was losing interest in receiver technology, which he now left to Ted Cope. Breaking into transmission, where Marconi had a monopoly, and studio equipment, dominated by EMI, was a new challenge.

By 1947 a new division of Pye known as Television Transmission (TVT) had fitted two cameras into a Humber shooting brake and a van to give the BBC part of its first proper outside broadcast unit. The next year TVT gave Copenhagen and Stockholm their first experience of television with demonstration broadcasts by local variety stars. By 1950 Pye was working with General Precision Laboratories in America to develop their innovative Image Orthicon camera that was later sold all over the world.

What the Stanley–Renwick campaign lacked was an authoritative figure whom the public would trust to *make* good television programmes rather than transmit and finance them. C.O. knew he had found the right person when Norman Collins resigned as the BBC's Controller of Television in October 1950. Explaining his decision in a letter to the *Times* Collins said the country had to decide whether television was to be allowed to develop 'on its own lines and by its own methods', or

Figure 7.4   Richard Murdoch being interviewed on a restricted pre-commercial television programme at Pye's own studio in Cambridge, 1951

whether it was to be merged into 'the colossus of sound broadcasting' and forced to knuckle under to a BBC that regarded the new medium with 'disinterest and sometimes open hostility'. C.O. at once telephoned Collins and offered to invest £25 000 in any television project he cared to start. The cultured ex-BBC man and the pugnacious salesman who thought he knew best were not natural partners; the only time the two had met was when C.O. went to see Collins at Alexandra Palace to lecture him on the poor quality of BBC programmes. They hit it off because C.O. did not claim to know how to make programmes and never interfered in Collins' domain. As for Collins, he matched Renwick in the patience and humour needed to put up with C.O.'s insults and rages. He even managed to reply sweetly to C.O. when in one terrible outburst he accused Collins of failing to show gratitude for his help.

C.O.'s energy and money and Collins' expertise gave birth to High-Definition Films, the first attempt to produce 'canned' film economically and quickly for television. C.O. had spotted the need when he decided BBC programmes were failing to entice the public to buy TV sets. Conventional films were too expensive and slow to make to satisfy television's huge appetite for material, while no one would readily watch the low-quality film taken direct from the cathode ray

Figure 7.5   *Humber shooting brake fitted with a television camera for outside broadcasts, 1947*

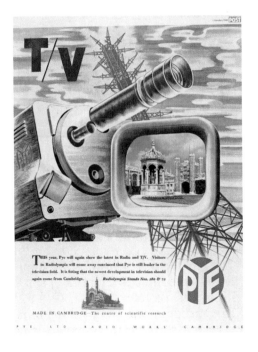

Figure 7.6

*Figure 7.7   Norman Collins, who resigned as BBC's Controller of Television in 1950
and who subsequently became prominent in commercial television*

tube. The technology of High-Definition Films, developed in the Pye
laboratories under B.J. Edwards' direction, called for High-Definition
television cameras made by Pye, and a fast but otherwise conventional
cinematograph camera. Recording a half-hour television programme in

one day was now a possibility. C.O. promised that when 'the first electronic film-recording studios in the world' began work in 1952, 'Great Britain, where television made its start, will go pioneering again'.

High-Definition Films started business in mid-1951 with an initial capital of £100 000 of which Pye put up a quarter, B.J. Edwards having assured the chronically cash-hungry board that the new company's shares could be 'realised at a later date at a considerable profit'. Headed by Collins, its directors included C.O., Renwick, London's leading literary agent A.D. Peters to look after artists' contracts, and Viscount Duncannon (later the Earl of Bessborough), a well-connected merchant banker at Benson Lonsdale.

Planning for High-Definition Films did not mean that C.O. gave up his own television campaign. In February 1951, just after the publication of the Beveridge report, Conservative MPs set up a Broadcasting Policy Committee, with C.O.'s employee Anthony Fell as its secretary, in which the majority were opposed to the BBC monopoly. Two months later C.O. put on a demonstration of sponsored television for Members of Parliament before the House debated Beveridge's findings. Pye set up a studio off Chancery Lane and installed receivers in a House of Commons committee room for a half-hour variety show compered by the BBC's Richard Dimbleby and including stars such as Jack Hulbert, Cicely Courtneidge and the TV chef Philip Harben. Examples of likely British commercials were shown between the acts, for C.O. was anxious to prove his point that British television advertising would not copy what were seen as the excesses of America.

The timing was better than he knew, for the autumn 1951 election that brought in a Conservative government would mark the start of a revolution in British broadcasting. Much was to be written later about a pressure group of Conservative MPs who promoted commercial television for their own financial interest. It was true the election brought in a new sort of Tory member who was more business-conscious and more suspicious of monopoly than many of the older members, but few of those on the Broadcasting Study Group would have any financial interest in the new commercial companies. As Asa Briggs pointed out, in 1951 Britain was poised to pass from austerity into a world of rising expectations. Many people could at last hope to move into new houses, and television was the perfect advertising medium for the consumer goods that would come with them.

No one seemed better placed than High-Definition Films to exploit the new opportunities. At this stage neither C.O. nor any other member of the High-Definition team was thinking of a separate commercial channel; their aim was to provide an hour or two's sponsored television a day to be shown over the BBC network in addition to its own programmes. But after Collins, Renwick and Duncannon saw the

Figure 7.8   The Earl of Bessborough, who as Viscount Duncannon emerged as one of the leading figures behind sponsored television in 1952

Conservative party chairman Lord Woolton in January 1952, Collins told C.O. there was a better chance than ever to 'upset' the BBC charter with 'an entirely independent television service', though this could come about only when the government relaxed restrictions on the steel needed to build new transmitter towers.

The problem, Collins told C.O., was persuading the public that a worthy alternative to the BBC existed. Since 'nothing less than a body representative of major national interests' would do, he suggested an alliance between Britain's leading film and electronic companies, possibly the Rank Organisation and Sir Alexander Korda from the former, with Pye, EMI and English Electric representing the latter. There was now 'extreme urgency' because the government was preparing a new broadcasting white paper, and a large group of Tory backbenchers were about to propose breaking the BBC monopoly. Once this became public there could be 'a silly rush [by] irresponsible people' to register 'mushroom television companies'.

With his salesman's instinct and his perception that television was above all a popular medium C.O. needed no encouragement to move fast. He was deaf to cries of alarm from the establishment about the dangers of commercial broadcasting, not least because he was convinced it was a force for the good. Speaking at the Dublin Spring show in 1951 he insisted Pye did not only see television as a commercial enterprise. 'It is a great art, that, educationally and socially, can bring something into the lives of millions of people, and that is very much more important than selling a piece of equipment.' He spelt out his attitude to the BBC in more detail in Pye's annual report of 1952. The war, he wrote, had brought home the power of the spoken word for both good and evil, and the election of a Labour government made 'many thinking people' in Britain wonder 'if we had tied a millstone round our own necks by having a monopoly broadcasting system which, if Socialism led to National Socialism or Communism, could be completely government-controlled'. The BBC had suffered a cut in income, to the detriment of programme quality: 'admittedly the [licence] fee is cheap but in America there is no fee at all and you can take your choice'.

It was natural to argue, C.O. went on, that 'a monopoly in broadcasting could be a threat to freedom and that competition would encourage better entertainment'. Nevertheless it only needed the government white paper to mention the possibility of competition to bring down a 'storm of abuse' on the radio industry and other so-called 'pressure groups'. C.O. challenged the assertion of the former director general of the BBC Lord Reith that there was nothing wrong with monopoly provided it was 'conditioned by responsibility, wisdom and courage'. Was not competition the most effective 'conditioning agent'? As to accusations that money, including Pye's, had influenced Parliament he countered that 'the only significance that money has in all this affair is the cost of running a television programme – that and that alone will decide how and when there is an alternative programme to the BBC television monopoly'.

This was C.O.'s strongest point, for the Labour government had proposed that the BBC borrow money on an unprecedented scale to extend its television service to the whole of Britain. Few realised at the time that C.O. and his colleagues were quite as much in need of cash as the BBC, although that did not stop them setting up their own television company. In July 1952 Norman Collins gave C.O. an account of a meeting with Lord de la Warr, the new Postmaster General. He had found de la Warr shell-shocked by the violence of the parliamentary debate on broadcasting in which the Labour deputy leader Herbert Morrison declared commercial television to be 'totally against the British temperament, the British way of life'. Collins described the Postmaster General as

> a country gentleman of peace and benevolence . . . loath to become the central figure of political controversy. However he is heartily sick of all the rumpus . . . and would like to find a way . . . which would satisfy the free enterprise without bringing the whole wrath of the Labour Party down on to his head.

De la Warr told Collins that steel was still in too short supply to allow commercial firms to put up their own transmitters within the next four years but, if Labour agreed, it might be possible for a company to use BBC transmitters to broadcast separate programmes of its own. This was what C.O. and Renwick wanted to hear, and when planning the launch of their company they first thought of calling it the Vision Broadcasting System, specifically to avoid the impression that it was planned as 'an alternative broadcasting corporation' to the BBC. In the end it was launched in August 1952 as ABDC, the Associated Broadcasting Development Company, with Collins as chairman and C.O., Renwick, Duncannon, the film-maker Sir Alexander Korda and the chairman of EMI Sir Alexander Aikman as directors. C.O. insisted that Aikman be included on the grounds that it would look 'extremely bad' if he was the only representative of the radio industry on the board. He had also decided he could dominate Aikman, whom he considered had 'little ability but [was] very easy to get on with'.

ABDC's delicate lobbying missions were usually carried out by Collins, Renwick and Duncannon, a tacit recognition that C.O.'s presence might do more harm than good. Apart from paying the company's still small running expenses, and rather infrequent attendance at its board meetings, he concentrated on the High-Definition project. High-Definition seemed essential to ABDC's success as a would-be programme company, but C.O. also hoped to sell or lease round the world the equipment Pye was making for it. The first of the new cameras was to be ready by the end of 1952, and films made for television by the new process the following year.

When ABDC applied for its broadcasting licence it was told nothing could be done until TAC had resolved the pressing matter of frequencies, At the urging of its backbenchers the new government had revived

Figure 7.9   Press reveals the creation of commercial television, 1952

TAC to advise the Postmaster General on frequency allocation for television and sound broadcasting including possible 'competitive television services'. Not surprisingly there were protests when the government appointed C.O. as one of the two members representing industry in a TAC otherwise dominated by civil servants and officials of the BBC. C.O. had been to see the deputy Postmaster General John Gammans, who was sympathetic to ending the BBC monopoly, to suggest he put Collins or Renwick on TAC. He was also in touch with John Profumo, a leader of the Conservative Broadcasting Study Group who was urging the government to make more use of TAC as long as it was not packed with BBC sympathisers.

No one was more shocked by C.O.'s appointment than his own industry, for it knew too well that he would represent only himself (and indeed C.O. withdrew from the Radio Industry Council the following year). *Wireless World* argued, correctly, that C.O. had 'pre-judged' the issue of AM versus FM broadcasting that TAC was also supposed to decide (discussed in the preceding chapter), and asked if C.O., as one of the founders of ABDC, proposed to 'advise the Postmaster General on the highly controversial issue of the conditions on which his own [television] company should be licensed?'. According to *Wireless World* the only solution was for C.O. to be 'allowed to resign [from TAC] with honour', a recommendation he ignored. (After joining TAC he did sell his shares in his advertising agency Arks, but only to one of the family trusts whose management he was able to influence.) His name also came up in the House of Commons when Ian Orr-Ewing congratulated the Post Office on putting someone with 'knowledge and enthusiasm' on TAC, while the former Labour Postmaster General Ness Edwards said C.O.'s appointment made the Post Office 'look like Tammany Hall', a reference to the notorious crony politics of old New York.

C.O. turned the attacks against him to his own use. When TAC's report on television was published in May 1953 he wrote to its chairman Admiral Daniel asking to be relieved of his committee member's obligation not to speak to the press 'because of the depths which have been reached by the campaign for state broadcasting as opposed to free enterprise'. Daniel, who handled C.O. well, understood this as a threat of resignation and persuaded him to do nothing until the government made its statement on television policy. Inevitably C.O. also added a 'reservation' to TAC's decision on television broadcasting frequencies. His quarrel with his colleagues was that they did not go far enough. TAC's recommendations covered the allocation of bandwidths necessary to contain increased television services, specifically that Bands IV and V be reserved for television and, eventually, all of Band III as well. C.O. pointed out that this did not solve the problem of how to satisfy users of mobile radio, most of them Pye Telecomm customers, who were the present occupiers of Band III. TAC had also ignored industry's (i.e. chiefly Pye's) demands that the Post Office hand over control of frequency allocations to an independent body similar to America's Federal Communications Commission.

The TAC report took no stand for or against commercial television as such, and was soon swamped by an emotional public debate. C.O. and Collins had lost their battle to make people think of a new television service as 'competitive' or 'independent' rather than commercial, and the prestige of the BBC had been further enhanced by its coverage of the Coronation, which earned it the accolade of 'best broadcasting service in the world' from the *News Chronicle*. Jules Thorn, one of the most powerful figures in the radio industry, had come back from the

United States openly hostile to commercial TV, while ABDC's own director Alexander Korda was warning that bad American-sponsored programmes might drive out good British ones. The Labour leader Clement Attlee threatened that if the Tories did bring in commercial television Labour 'would have to alter it when we get back to power'.

It was not surprising that in such an atmosphere C.O. ignored Daniel's counsel of restraint. That summer Pye's annual report contained a long article under the title 'Commercial television – for or against' which, though unsigned as usual, summed up his thoughts. It began by laying out the two main arguments against commercial TV: that it would not match the BBC's 'high standards', and that 'the public will be subjected . . . to a continual stream of advertising propaganda which, though apparently perfectly innocuous [when they appear] in the press, will be utterly objectionable on television'. ABDC would ensure quality by appointing a retired BBC mandarin as 'chief programme adviser'; as for advertisements, there was no reason why a British commercial channel should not respect British taste. And it was nonsense for the critics of commercial TV to argue that a housewife would buy an unsuitable product just because she had seen it advertised on a good programme – 'as Churchill said during the war, what sort of people do they think we are?'.

C.O. used the article to list, and ridicule, commercial television's chief opponents. The press feared losing advertising to a competitor; newspapers that had applied for licences to set up commercial stations did so without enthusiasm and probably hoped 'the necessity will not arise'. (The *Daily Mirror's* Cecil King had made friendly noises to ABDC, but C.O. warned Collins that 'all the assurances in the world from a newspaper proprietor are quite worthless until they are signed up'. King was heard to say of C.O. that he had 'the mind of a small businessman'.)

The film industry feared change because it was in trouble, and did not understand that only 'cheaper production costs for better films, and participation in commercial television' could save it. Then came 'the commissars of the intelligentsia . . . who think that they know what is good for us, that our lives should be planned and that they should do the planning'. These were the 'perfect props of the Welfare State', the people who have 'never built anything – nor managed anything – nor made anything – nor sold anything'. There was also the Labour Party, which on this issue showed itself to be the party of reaction while Conservatives, though split, were the progressives. As for the BBC, it was bound to oppose any challenge to its monopoly, even though 'the biggest argument in favour of commercial television' was that it would 'disturb the BBC'.

Finally he defended the supporters of commercial television from the charge that they were 'pressure groups'. There was nothing sinister in the fact that 'industry believes in profits and believes profits are related to turnover'. Commercial television would bring more, and more popular, programmes, more people would buy TV sets, prices would fall, employment would rise, and British manufacturers would be able to compete in the world market against the far bigger industry of America.

> This country's welfare was never built on the Welfare State . . . The men and women who matter are not the ones who play safe and want much for nothing and wait for a pension. Britain, perhaps more than ever before, needs a hard competitive commerce. There is nothing wrong with the world "commercial".

The formation of the National Television Council in June 1953 gave C.O. the chance to abandon what remained of his TAC-imposed discretion. An all-party group pledged to resist commercial television, the Council was directed by the broadcaster and Labour MP Christopher Mayhew whose BBC sympathies earned him the nickname of 'the Honourable Member for Lime Grove'. Council members included the Conservative Lord Halifax and an eminent Liberal, Lady Violet Bonham Carter, but it was above all an alliance of the high-minded. Bertrand Russell, the novelist E.M. Forster and the historian Harold Nicolson joined with university vice-chancellors and church leaders who, according to the *Daily Sketch*, feared that once advertising came to broadcasting 'nothing [would be] sacred'.

These were the people C.O. mocked as 'Twentieth Century culture gospellers' afflicted with the 'pious nonsense of totalitarian minds', and their coming together offered an irresistible target. Mayhew, unaware of the ferocity of the coming battle, predicted a 'walkover' for his side. Summer opinion polls showed most people still did not want commercial TV. The Labour Party and the learned professions solidly opposed it, and victory seemed assured if Mayhew could win over enough members of a divided Tory Party. American television's use of a chimpanzee called J. Fred Muggs in commercials shown in breaks between a film of the Coronation further strengthened the antis: Mayhew won a well-publicised debate on commercial television at the Oxford Union holding a chimpanzee by the hand while he spoke.

Mayhew thought the National Television Council was 'like much public service broadcasting, . . . weighty, honest, public-spirited and poor'. Its members felt contempt for the commercial lobby that had 'all the merits and defects of commercial television . . . populist, mendacious, mercenary and rich'. The embarrassing truth was that ABDC still had a nominal capital of only £100, and did not know where to raise more. It was Renwick's job to find the £1 million they thought they needed to start operating. Pye, cash-short as ever, did not have resources on that scale, and few people then saw commercial television as the licence to print money the newspaper proprietor Lord Thompson later called it.

Within a month of the appearance of the National Television Council C.O. chaired a dinner at the St Stephen's Club in Westminster, chosen because many of those present were Tory MPs (among them Anthony Fell, Kenneth Lewis and C.O.'s clever Ulster friend Willy Orr) who might need to get back to the House to vote. When the meal was over C.O. made a short statement about the formation of a Popular Television Association, and asked Renwick if he thought there would be money to finance it. Renwick stood up and pledged £20 000 which earned him cheers and a joke in poor taste from Fell who proposed a toast to 'Sir Money Bags'. The joke was also inaccurate for Renwick, well-off but not rich, was a raiser, rather than a giver, of money. The sum he pledged that night almost certainly came from C.O. and Pye.

The dinner ended with a request to Norman Collins to approach Lord Derby to be the Association's president. Other celebrities brought in by Collins included the cricketer Alec Bedser, the actor Rex Harrison, and the historian A.J.P. Taylor, who welcomed the prospect of ending the BBC monopoly as 'the biggest knock respectability has taken in my time'. The presence of Lord Derby and other aristocrats did not spoil the impression of a more popular body than Mayhew's Council. The Association's publicity pursued two themes: monopoly was evil, and advertising on British television would observe British standards of taste and avoid American 'excesses'. No sooner had the campaign begun than C.O. was complaining that it lacked energy, and arranged for Ronald Simms, a public relations consultant who later moved to Conservative Central Office, to take it in hand. C.O. himself became part of the debate and a Tory backbencher complained it was 'impossible to discuss the subject without hearing his name mentioned'.

Meanwhile ABDC was still arguing that it would be possible to put out commercial television over publicly owned transmitters, and earlier in the year had offered to pay the BBC an annual £4 million to rent the BBC's transmitters for three hours a day (where it would find such money was not explained). ABDC talked of partnership with a national newspaper, though Renwick was a long way from striking any deal in Fleet Street. The scheme was forgotten when the government's television white paper published in October 1953 proposed setting up a public corporation to own television transmitters and issue licences to the commercial companies that used them.

Some called it a victory for the commercial lobby. C.O. and those like him thought nothing of the kind, for they saw the creation of a new public corporation to regulate television companies – the nub of the government's plan – as yet another attempt to put chains on independent broadcasting. In the Commons Pye's Anthony Fell called the white paper the most depressing document he had ever read, and melodramatically declared the evils of freedom to be always preferable to the evils of 'authoritarian dictation'. Those were C.O.'s sentiments too, and they explain the occasional outbursts of protest, and almost of anguish, with which he would disturb his colleagues as they pursued their more flexible idea of commercial television in the years ahead.

Asa Briggs described the March 1954 Television Bill as including 'far more don'ts than dos', and ABDC now found itself fighting to prevent the Conservative government deviating from what it understood as the true principles of free enterprise. Robert Renwick consulted with C.O. before sending their objections to Conservative Central Office, and his insistence on playing by his own rules found reflection in Renwick's argument that it would be impossible to operate a programme service under the terms of a Bill 'so fundamentally wrong in spirit that [there is] no point in suggesting alterations'. Renwick also warned that the proposed legislation would cost the Conservatives the next election 'on the issue of free enterprise as represented in the particular instance of commercial television'. For Renwick these hard words were a tactic to win amendments to the Bill, but C.O. meant every one of them.

The Television Bill became law in July 1954 and in August the Independent Television Authority (ITA), the government-appointed watchdog that C.O. wanted nothing to do with, began work under the chairmanship of Kenneth Clark, eminent art historian, former director of the National Gallery and apparently just the sort of representative of high culture C.O. distrusted. He did not know that when Clark was taken to lunch in the Athenaeum shortly after his appointment, the bishops and academics who populated the club booed him as a traitor to culture. Nor did C.O. know that Clark accepted the chairmanship of ITA in the hope that commercial television would add what he called 'vital vulgarity' to British broadcasting.

Clark was as little taken with C.O., and was amazed when Norman Collins remarked, after C.O. had thrown one of his fits of temper, that he was 'not to be judged as ordinary men are judged because Charles [is] a genius'. Asked what he meant by 'genius', Collins defined it as C.O.'s ability 'to energise any situation whatsoever'. Certainly it was energy, not money, that was keeping ABDC going. In spite of their dislike of the new Act its directors submitted an application for a licence to provide commercial television seven days a week to the London area, promising programmes that included news, drama, religion, light entertainment and outside broadcasts for sport. They claimed that the necessary capital of £750 000 was already 'available', but since losses were inevitable in the first two years they asked for an eight-year contract to guarantee eventual profit. Even if £750 000 was 'available' it was not enough. Collins reported to C.O. a conversation with Paul Adorian of Rediffusion, another would-be programme contractor that was spoken of as a rival to ABDC for the London station. Adorian boasted that his company already had funds of £2 million, and told Collins that ABDC's weakness was its lack of money. Collins disagreed, but admitted to C.O. that he could only summon up 'a mood of controlled optimism'.

In October the ITA sent out a questionnaire to would-be contractors. Drafting ABDC's reply Collins described C.O. as 'a leader of the television industry and since 1935 . . . identified with every movement concerned with the expansion of television'. He pointed out that C.O. and Aikman could assure through their companies all the equipment ABDC needed to take to the air and claimed, accurately, that as a former Controller of BBC Television he was the only person outside the Corporation with proven ability in the medium. But he became vague when describing ABDC's plan to raise finance 'through various countries with which members of the board are associated and from other Sterling sources'.

Collins sent a copy of the questionnaire to C.O., who did not appreciate this latest evidence of nannyish control and scrawled on it, 'I don't like this'. He then wrote his own mock answers whose truculence suggest his anger at the way competitive television was taking shape.

'Question: Who will be your directors?
C.O.: I know of none.
Question: Are you aware your company must be under the control of qualified
   persons?
C.O.: What is a qualified person?
Question: What will be your financial resources?
C.O.: Unknown.
Question: How do you think news should be handled?
C.O.: Quite differently from the BBC.
Question: Would you be prepared to combine with other contractors if [ITA]
   wished?
C.O.: No.'

The following month the ITA refused ABDC's request to broadcast seven days a week in London, but offered it the London weekend slot and five weekdays in the Midlands. C.O. at once wrote to Renwick that he wished to resign from ABDC:

I have been disgusted by the political trickery that has gone on in this affair . . . The principles I have always supported of honest free enterprise must at no time be trammelled by behind-the-scenes conditions. I therefore feel that as I would have to be outspoken about these matters, I might well be an embarrassment [to ABDC. I am resigning] in order that you can do what you want without any feelings of restraint that may be created by my presence.

This was C.O. being what Norman Collins called 'very difficult' rather than a serious threat to leave. Perhaps it also hid resentment of his more nimble colleagues. Collins knew everyone who mattered in television. He did not demonise Kenneth Clark as C.O. did, and was an old friend of ITA's first director general Sir Robert Fraser, whom C.O. could never bring himself to trust. Renwick was a realist for whom compromise was as necessary as crossing the street. By the end of 1954 C.O. knew he was not going to get the sort of truly free enterprise television he had dreamed of, and the bitterness remained long after the threat of resignation was forgotten.

It was plain enough when he drafted his usual lecture to the nation in the following year's Pye annual report. The tirade, titled 'The Nursery Governess', was touched off by two statements. The BBC's director general Sir Ian Jacob had said there was 'too much television and that it destroyed silence'. And the Postmaster General had announced that the BBC and ITA agreed there should be only 35 hours of weekday television, and a complete shutdown each day from 6 to 7 pm. C.O. thought this ridiculous. 'The obvious way, one would have thought, to ensure silence from broadcasting, would [be] to switch off the set.' What outraged him most was the attitudes behind such pronouncements. The hourly evening shutdown was said to be necessary to 'assist parents in getting their children to bed ... [but] what if later ... it were decided that people should take a walk in the fresh air at night before retiring, rather than go straight from their little home theatre to bed?'. Would there be a new 'silent period' after 10 pm?

> Whether deliberately or not, the choice of phrases that was used to convey these ... arrangements to the public was more typical of a benev-olent Victorian parent who believed in the certain wickedness of his children than of a group of democratically appointed public servants.

His scorn reached a peak when he came to the ITA spokesman who had declared it a 'social *necessity* to impose restrictions' if the nation's 'home life [was not to be] greatly upset'.

> When we read this we were immediately reminded of a far finer spokesman speaking to the House of Common 172 years ago. *"Neces-sity"*, said William Pitt, "is the plea for every infringement of human freedom. It is the argument of tyrants; it is the creed of slaves.

The most pressing problem was not the ubiquitous British 'nursery governess' but ABDC's continuing lack of cash. It was the first pro-gramme company to set up and everyone, including the ITA, saw it as a natural choice for one of the first contracts. But it was impossible to award a contract to a company that might not have the money to pro-vide a service, and ITA made it even harder for ABDC to raise funds by offering it what seemed an unattractive mix of London and Midlands programmes.

C.O. attended the ABDC board meeting on 27 October that accepted the ITA offer. They had little choice for, as the Authority explained, it was already treating the company as a special case on 'the financial side' by accepting its pledge that each board member would raise £250 000 (which was quite unrealistic) and that it would also bring in a newspaper as partner. ABDC had mentioned the *Mirror* and Odhams Press but within days both dropped out and were replaced by the *News Chronicle*. This made C.O. uneasy. Cecil King, the only press magnate to show interest in ABDC the year before, had made it plain that if he gave money he wanted control. The prospect of losing even partial control to an outsider horrified C.O., and he insisted ABDC require that anyone 'putting up money should sign some sort of document such as an informal prospectus to which they must agree'.

The *Chronicle* dropped out too, Collins complaining to C.O. that its attitude had been 'niggling and mischievous', and ABDC was back on the financial tightrope. This may not have made C.O. dizzy – he had mastered the skill early in life – but it was an unusually dangerous sport given the controversy surrounding commercial television's creation. Much to C.O.'s annoyance the ITA persuaded ABDC to bring in two Birmingham newspaper groups and the Independent Television Programme Company (ITP) of the show business entrepreneurs Prince Littler, Val Parnell and Lew Grade, whose own application for a programme licence was turned down because they had little experience outside British light entertainment (where they had a near-monopoly). Clark was ready to accept them as ABDC's partner but he would not countenance a further ABDC proposal to bring in two Conservative newspapers, the *News of the World* and the *Daily Express*. Conservative papers were already involved in other programme companies and to allow more would have destroyed public confidence in commercial television before it had begun.

In February 1955 ABDC and the showmen were brought together formally in ABC, the Associated Broadcasting Company. Pye's investment was £170 000, while C.O. held at least 50 000 of the one shilling deferred shares. High-Definition Films, which had demonstrated filmed programmes to the ITA as early as September 1954, and ITP's control of big London theatres and entertainment stars gave ABC a head start in popular programme making, and at the new company's first board meeting John Stanley, standing in for his father, pledged that Pye would 'give its service without charge on technical matters' until ABC was in profit. Nevertheless the new company was an uncomfortable marriage. C.O. might have been expected to have got on well with Parnell and Grade, whose popular tastes he shared (not for nothing was Grade sometimes known as 'Low' Grade), but his dominating manner and eagerness for controversy unnerved them. There was a typically difficult moment when C.O. made a public speech arguing that sponsorship was a better way to finance television than commercials, and then claimed that ABC's decision in favour of commercials was taken in his absence. And his instant dislike of Harry Alan Towers, the director put in by Grade and Parnell to run ABC's programme production, would bring conflict into almost every board meeting.

The new commercial service was planned to start in London in September 1955, and in the Midlands the following December. Limited transmission possibilities forced ITA to split each station between two programme companies, an arrangement Pye's engineers were convinced, wrongly it turned out, was unworkable. ABC was to share a transmitter in London with Associated Rediffusion, which had the weekday broadcasts, and in the Midlands with the television wing of Associated British Cinemas, which in this case had the weekend slot. C.O. bid for the contract to build ITA's Midlands transmitters. By this time Pye was exporting the full range of television equipment, in 1954 pulling off a high-risk coup by demonstrating a television station and transmitter in Baghdad and then selling it to the Iraqi government. But the Baghdad transmitter, though only 500 W, was the most powerful television transmitter Pye had built; ITA's requirement was for three Band III transmitters, two of 5 kW and one of 20 kW, a task of a quite different order. B.J. Edwards and his transmission expert Jim Bennett persuaded a not very well briefed Post Office official over lunch that Pye would have no problem with them, though C.O. admitted to the Pye board that the transmitter equipment would be both difficult and unprofitable to make. Time was short – the smaller transmitter was supposed to be ready by January 1955, and the larger by June – and for once C.O. seemed nervous, emphasising to his colleagues that he had 'given his personal undertaking that the company would fulfil the contract'. At the same time Pye took on large orders for studio equipment and cameras for ABC, for Associated Rediffusion and Independent Television News. B.J. Edwards was known for his dislike of any sort of planning, but in this case he set tough production schedules that called for overtime and weekend work. He did not, though, provide close supervision and, apart from an occasional enquiry of 'How's it going, boy?' to Jim Bennett, left all technical matters to him.

A veteran of Pye's frantic wartime operations, Bennett was used to working in this way and thrived on it, but John Stanley was shocked when he checked on the transmitter team's progress in February 1955. In a long letter to C.O., away on his usual winter travels, he warned that though the completion date had already been put back the 5 kW transmitter would not be ready by the revised target of 1 September, and that Bennett and his team would be lucky to complete the more powerful one a year after that. Worse still, news of Pye's difficulties

Figure 7.10   Erection of Band III aerial in Cambridge, 1954

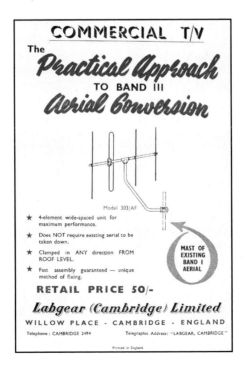

Figure 7.11   Subsidiary company of Pye releases aerials for Band III, 1955

Figure 7.12   Family in Baghdad watching a Pye television, 1956

had got out, and at a meeting of the Conservative 1922 Committee the Postmaster General, already under attack from backbenchers for being slow in getting commercial television on the air, 'had laid the blame fairly and squarely on Pye'. Edwards was on holiday but John called a meeting of other senior executives which decided that 'the engineering department have not the slightest idea when these transmitters will be ready'. John warned his father there could be a 'vicious attempt' to discredit Pye, and that 'one or two people in the Post Office' might try to use these difficulties to have him removed from TAC.

What John saw was perhaps not so different from the way Pye worked throughout the war, a state of creative confusion that suited Edwards and C.O. equally well. But to John Stanley's eyes Bennett, the man on whom most depended, was 'close to breakdown', though Bennett himself did not think so. Far more serious was John's suspicion that the real cause of the trouble was B.J. Edwards. How, he asked C.O., should he deal with 'the problem of Edwards' and his 'present mood' which threatened dire consequences? 'At this juncture any internal strife in our company would be dreadful, but it must be faced that there is a very real danger of that.'

C.O. had his own worries about Edwards, chiefly that his projects were costing too much money. He liked the idea of Pye getting into the complete television chain because of the opportunities it brought to promote the company, but he had to take into account that the transmission business never made money for him (nor did High-Definition Films whose losses he had to write off after the American Ampex Corporation invented the better technology of recording on magnetic tape). John was worrying about the effect of Edwards' increasingly erratic private life on his work, but this was something C.O. would not discuss. Traditional respect for privacy in personal affairs and uneasiness in intimate contacts encouraged him to believe Pye executives led irreproachable lives.

Preparations for ABC's first London broadcast in September 1955 turned C.O.'s dislike of Harry Alan Towers into an open feud. C.O. thought Pye should get the entire ABC contract for equipment in London, and warned Towers that he would consider anything less 'an extremely unfriendly act'. He was particularly angered by signs that Towers wanted Associated Rediffusion to share ABC's London studios, an arrangement C.O. had always opposed. Towers was unrepentant, telling C.O. that while he regarded Pye as 'best friends and most favoured nation', the Littler-Parnell-Grade team had always thought ABC would only be able to make profits by sharing facilities with the other London company.

C.O. denied knowing anything of this decision to become what he called a 'cap-in-hand' tenant of Rediffusion. 'I don't agree it now and I don't intend to be a party to it, so if you don't like this, it is just too bad'. As for Towers' insistence on his right to procure equipment from other companies '[it seems] you do not appreciate that the Pye company have been largely responsible for getting commercial television at all in this country'. Towers began to needle C.O. about the late delivery of equipment for the ABC control studio in Foley Street, for its TV studio at the Wood Green Empire and ITN's Kingsway studio. Six weeks before London transmission was due to begin Towers wrote to all ABC directors that he was 'more disturbed than ever' by Pye's slowness, and amazed that the company was not working over the August holiday to complete it. C.O. cabled Towers accusing him of panicking. Towers replied it was not he but Pye's crew in London who were 'so depressing about the whole situation'. A telegram from Edwards promising all would be ready for opening night on 22 September made matters worse. Towers pointed out they had to be ready well before that date to carry out test runs, and twisted the knife by adding that Pye needed 'a sense of urgency instilled . . . from the highest level'. Although Towers did not understand how Pye worked (how could he?) he may have had an inkling that the company was overburdened.

London's first 'competitive' television programme did go on air as planned on 22 September. The first commercials for toothpaste, drinking chocolate and margarine were, as C.O. had promised, inoffensive to British taste, an American newspaper reporter calling them 'painless by American standards'. But Towers had been right to worry, for the first broadcast was only possible thanks to Pye's talent for improvisation. The master control room in Foley Street was not finished in time, and was in working order only because the Pye engineers found the switches, resistances and other components they still lacked in second-hand shops in the Edgware Road. The perfect pictures on the screen were achieved by what engineers called a 'temporary lash-up'.

The launch was marred by a dispute, to which C.O. contributed his own brand of obstruction, over ABC's right to its name. The Associated British Picture Corporation claimed the abbreviation ABC for its subsidiary Associated British Cinemas (Television) which had won programme contracts for the Midlands and the North. C.O. thought the company had already made too many concessions and decided to stiffen the backbone of his show business colleagues. '[We have] already done a great deal of appeasing in the last few months and, in my opinion, the principle of appeasement is not only always wrong, but it makes it impossible to do good business with people who have lost their respect for you.' He argued that if the board did agree to a name change it should demand £750 000 compensation because 'the effort which I personally put into ABC before you came into the picture was worth more than a quarter million'. The name was lost – it is hard to see how it could have been otherwise – but Associated British Pictures did agree to share some capital costs.

C.O. went on to complain about their new name, Associated Television. He said its abbreviation as ATV reminded him of the wartime ATV, the Armoured Transport Vehicle, which C.O. claimed had been such 'a flop' that it was in bad taste for a new broadcasting company to use the same initials. There were several sorts of armoured transport vehicle in the war, and light armour did make them vulnerable, but they were scarcely the detested machines that C.O. described. Angry over the way commercial television was developing, he seems to have been making trouble for trouble's sake, and showed no surprise when his protest was ignored.

The New Year brought him the satisfaction of seeing his enemy Harry Alan Towers forced to resign for taking unauthorised commissions. One of Towers' last impertinences was to accuse Pye of threatening to remove equipment from the ATV studios unless it received prompt payment. C.O. denied making the threat, but displayed all his capriciousness in a complaint of his own to Norman Collins. Claiming ATV had told Pye it did not have enough money to pay its bill, he said it was 'completely shattering to think of a company your size only a few months in existence not having enough money to buy this equipment'. To write as though he himself had nothing to do with the problems of a company of which he was a founding director was unusual cheek even for him.

The Towers affair, with its hint of more difficulties at a company known to be short of cash, came at the worst possible time for ATV. The opponents of commercial television were using the supposedly poor quality of the first London programmes as ammunition for further attacks and ATV needed all the political support it could get. Fearing MPs such as Fell and Orr-Ewing were too obviously associated with ATV the board proposed that C.O., Collins and Renwick should give a lunch at the Commons to lobby a wider group of Tories. C.O. agreed, but with the caveat that, if he spoke to the MPs, he would defend only the 'principle' of commercial television, and not 'commercial television as we have seen it [in Britain] up to now'. He remained convinced Britain was wrong to have chosen commercials over sponsorship, and would not hide his disappointment even to help win much-needed friends for ATV.

The chief danger, though, remained the old one: lack of money. At one crisis meeting John Stanley, standing in for his father, was pacing up and down the boardroom floor complaining 'money is being spent like water and we will go bust' when coins slipped through a hole in his trouser pocket and fell on the floor. Everyone laughed, but by the end of 1955 the shortage of funds was causing speculation in the press and nervousness among shareholders. By the start of 1956 cash requirements reached £2.25 million (£500 000 was needed to launch the Midlands station in February). ATV asked the ITA to extend credit for £200 000 rent due over the next 6 months, and was shocked when this was refused.

C.O. was also under pressure at Pye from the deadline to complete the transmitters for the Midlands. The opening date had been put back to 17 February and work was proceeding, in Bennett's words, 'with great cost and teeth gnashing'. The Pye team had a lot of keenness, limited experience and never enough money. They worked several miles outside Lichfield in a hut under the new transmission tower that had no proper drawing office or any of the other facilities that engineers from Marconi, Britain's leader in transmission technology, would have expected. Dust and rubble from the construction of the tower contaminated air that needed to be clean for reliable transmitter operation. Bennett said later he always knew he could do the job even if others at Pye, though not C.O., 'were jumping up and down like mad', but it was a close-run thing. The first programme, a live broadcast from the London Palladium, was planned to go on air at 8 pm. The Pye engineers, confident everything was ready, were at their hotel eating an early evening meal when the ITA maintenance engineer at the tower telephoned to say the transmitter was not working. They hurried back and the Palladium held the show while they repaired the fault. Commercial television came to the Midlands 30 minutes late.

Figure 7.13   Popular ATV programmes, 1958: Clint Walker Stars in 'Cheyenne'. ATV was the first to screen an hour-long Western for week-end viewing

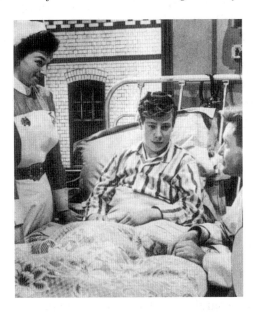

Figure 7.14   Popular ATV programmes, 1958: 'Emergency-Ward 10'. The twice weekly hospital series which won a TV Oscar from the Guild of Television Producers and Directors

Figure 7.15   Popular ATV programmes, 1958: Tommy Trinder and girls in Sunday Night at the London Palladium

Pye's transmitter troubles were not over. The following month the ITA warned ATV they were having difficulty bringing the Midlands transmitter up to full power. This meant transmission depended on the 5 kW transmitter, and if that failed ATV would go off the air. The ITA had accepted this risk because it believed Pye's assurance that the 20 kW transmitter would be in service by midsummer at the latest. This was now unlikely and 'the trouble is entirely concentrated in the delivery of Pye transmitters'. John Stanley's alarm in early 1955 was beginning to look like an accurate assessment of Pye's unreadiness to take on a contract of this size.

By March 1956 the problem of the new transmitters was overshadowed by a new financial crisis. The National Provincial Bank had to extend ATV's overdraft facility in exchange for a general charge on its assets, and the company could not afford to take part in the Radio Show later that year. Losses in the 14 months from February 1955 to April 1956 came to over £600 000 and were now running at a rate of £1 million a year. C.O.'s joke answer to the ITA questionnaire – financial resources? 'Unknown' – seemed embarrassingly close to the truth.

In April the ITA allowed ATV to keep afloat by bringing in the *Daily Mirror* as a major shareholder. Cecil King had always liked the idea of commercial television but stayed out at first, telling colleagues 'let us keep our powder dry for the re-financing that will be needed in a year or so's time'. It was a shrewd approach, quite different from C.O.'s, which was to plunge in before anyone else and leave the finances to take care of themselves, and not surprisingly he was an object of much interest to ATV's new directors from the *Mirror*. The paper's editor Hugh Cudlipp and the Mirror Group's legal adviser Ellis Birk were shrewd men with a wider experience of the world than Parnell, Littler and Grade, and perhaps better equipped to assess C.O. at the height of his powers. When the men from the *Mirror* joined the board, C.O. was proposed as chairman of the enlarged company. He refused because of his interest in television manufacturing, and Renwick was picked instead. It was a much better choice, as Cudlipp and Birk soon discovered. The *Mirror's* columnist and former Arks employee William Connor warned his editor, 'Watch it. C.O. Stanley is as clever as a cageful of monkeys', but Cudlipp took to his new colleague. At board meetings he watched fascinated as a smile came and went on the face of

> this elderly Irish Leprechaun [giving the impression] he knew a thing or three beyond the ken of anyone else around the table, which was often true. The smile was mischievous rather than conspiratorial as he timed the effective moment to intervene with a view that was occasionally eccentric but always shrewd, his voice rising to a crescendo and his right hand slapping the table, temporarily obliterating any opposition until he had said his piece.

Ellis Birk came to a more measured appraisal. For him C.O. was 'one of the world's great disagreers – there was a really quite extraordinary perversity about him'. But Birk also thought he had a 'genius ... for the development of ideas ... [and] at seeing an opportunity and fighting

Figure 7.16   Celebrating the success of commercial television, for which C.O. Stanley had vigorously campaigned. Cartoon in Pye's Annual report, 1957

Figure 7.17   C.O. Stanley among the 'money men' of ITV, 1961

Figure 7.18   ATV's success marked by the Beatles, c.1964

Figure 7.19   C.O. Stanley and Norman Collins at an ATV dinner, 1970

to realise [it]'. When ATV had a battle to fight, C.O. was 'unstoppable' in organising support. By contrast, whenever he became involved in ATV's day-to-day management he was 'a menace'. Birk noticed how C.O. could seduce with charm, but be 'bullying and tiresome and relentless with people who didn't share his vision'. He did not care for C.O.'s noisy lectures to the board on hobby-horses such as the iniquity of the Post Office, but put up with them once he learned from Robert Renwick that the only way to deal with C.O. in combative mood was to laugh at him.

By 1958 Pye's £170 000 investment in ATV was worth £2 million. C.O.'s great adventure had paid off, and he would remain a director of the company he helped found until his retirement in 1975. When his fellow directors proposed a farewell lunch he refused, telling Norman Collins he did not need a party in his honour because 'what I did for television or ATV gave me intense intellectual satisfaction' which would be 'destroyed' by a formal ceremony of thanks. Collins recognised this as an intentionally awkward display of principle, and after more negotiations the lunch took place at Christmas with turtle soup, turkey, mince pies and Pichon Longueville 1967. But there was truth in what C.O. said to Collins. Had he not taken such satisfaction in the battle, he would not have been such a formidable fighter.

Sources

COS files: 2/2 (TV transmission, B.J. Edwards), 2/17 (Pye annual reports), 5/1, 6/1 (speeches).

Simoco boxes: Box 2 (High-Definition Films), Box 4 (ABDC; ATV including board meetings 1952–1956 and management meetings; TAC), Box 5 (High-Definition Films 1951–1952; ABDC 1952–1953).

Pye main board minutes.

Interviews: Lord Ashburton, James Bennett (transmitters), Ellis Birk (ATV), Sir John Clark, Mike Cosgrove, Jo Fletcher, Fred Keys, Anthony Lucas (ATV), Bill Pannell, Geoff Peel, Lord Renwick, Peter Threlfall, Daphne Whitmore, Michael Worsley.

Michael Bell interviews: James Bennett, Norman Collins, Les Germany, John Stanley.

A–Z files: James Bennett, Lord Cudlipp, Richard Ellis.

Books: Bell, *ibid.*; Bessborough, *Return to the Forest*; Briggs, *ibid.*, Vol. IV: Clark, *The Other Half*: Geddes and Bussey, *ibid.*; Mayhew, *Time to Explain*; Sendall, *Independent Television in Britain*, Vol. I.

Chapter 8

West Briton

In most summers after the war, usually in July, the Stanley Rolls Royce set off from Lowndes Place on the long drive to the Irish ferry and two months' holiday at Lisselan, C.O.'s house in County Cork. He sat in front with the chauffeur; Velma, a fur rug over her knees and accompanied by her maid-cum-cook Mary Sullivan, in the back. Velma dressed up to travel, putting on a hat even for these summer journeys and, if the weather called for it, a full-length fur coat. Good taste, expensive clothes and what C.O.'s young nieces thought of as 'amazing' legs made her a strikingly elegant figure. C.O.'s interest in his own clothes diminished as his success grew, but he took care to dress properly when travelling with Velma.

Her manner and his eminence ensured a royal reception when they arrived at Fishguard to board the ferry *Innisfallen*. They always took 'A' cabin; when told on one occasion that it was not available Velma declared the boat would not sail until she got her usual berth, which in the end she did. She spent the weeks before their departure collecting antique furniture, wallpapers and fabrics for Lisselan as well as foodstuffs that were then unobtainable in Ireland. This cargo was sent ahead to Fishguard and on arrival there Velma went to the quayside to supervise its loading. If a new port official questioned her right to have so much baggage transported free she told him she had been doing it for years and was the ferry's best customer. Irish demands for import duty were turned aside by the same mix of hauteur and charm. It was said that County Cork had never seen such beautiful clothes, but it helped that C.O. arranged for the ferry company's manager, and anyone else who was useful in their Irish comings and goings, to have shares in Pye and ATV.

Ireland had played a part in C.O.'s wooing of his wife. In one of his early letters he told Velma he wanted them to sail together into

Figure 8.1   Lisselan, Clonakilty, Co. Cork, Ireland, c.1930

Dublin at dawn: 'It isn't Venice, it isn't the East, it's much more fantastic.' Most of all he wanted to show her Lisselan, the house he had bought in 1929 after launching Pye on the Stock Exchange. He saw it first on a grey, damp day and the garden, screened by trees, was invisible as he drove up to the house. The place revealed its magic only when he went into the library where a fire was burning and he looked from its bow window over the terraces and lawns that fell away to the Argideen River running through the valley on whose sides the garden was laid out. Designed in the style of the 19th century gardener William Robinson, it was rich in old azaleas and rhododendrons, while the ivy-covered house with its grey mansard roofs and pinnacled tower had the appearance of a neat French chateau. Anything less like John and Louisa Stanley's home on Cappoquin's Main Street was difficult to imagine. He loved the garden, and Velma came to share his passion, developing into a formidable plantswoman who bombarded Cambridge University's Botanic Gardens with enquiries and requests.

In another of his courting letters C.O. imagined what it would be like to arrive at Lisselan at daybreak 'when the dew [is] all over the grass and [we can] smell all those smells that earthy places make and tramp around all the day and leave in the evening when the moon [is] up'. But Velma never developed a taste for this wilder side of Ireland. Tramping round was not her style any more than the fishing and sailing that had always been part of C.O.'s Irish life; nor was getting up early to see the dawn, for she was a notorious late riser.

Figure 8.2   C.O. Stanley and his son, John, canoeing at Lisselan, c.1939

Lisselan was C.O.'s domain; hers was Sainsfoins and above all Lowndes Place. She had decorated the latter in the most formal style with bronzes and antiques, and allowed none of the clutter of ordinary life to spoil her immaculate reception rooms. It was here that she organised C.O.'s entertaining, handling important guests with greater skill than her husband, who was inclined to lecture people rather than talk to them. Velma had little time for women; she liked men and was a good raconteur, telling risqué stories that she punctuated with a remarkably dirty laugh. She would embarrass C.O. by refusing to leave the table at the end of dinner as women were then supposed to do, and thought nothing of hitching up her skirt to warm herself better in front of a fire.

At Lisselan there *was* clutter on the tables, and though Velma redecorated the house and bought books by the yard for the library, C.O. imposed limits on her activities. He also refused to give up the Irish food he had eaten since childhood, and Lisselan's Sunday lunch never varied from boiled gammon and roast chicken accompanied by waterlogged cabbage and floury potatoes boiled in their skins. Her London dinner parties often ended with bridge, a game she played with passion, but in Ireland Velma's entertaining did not always run to plan. One evening at Lisselan when the women were waiting at the card tables set up in the drawing room C.O. decided not to join them. Bent double so as not to be seen from the house he led his male guests down to the Argideen to fish. The most boisterous evenings ended with the rolling back of the carpet in the library and C.O. dancing Irish jigs.

They fought often and furiously, but never for long. She invariably kept people waiting and one night when she had taken even longer than usual dressing for a party C.O. got bored and went down to his river for a few minutes' fishing. He lost track of time, then slipped and fell into the water, a crime made worse because he was wearing a dinner jacket she had just had made for him. He returned to the house to find waiting guests and an incandescent Velma. They were a match for each other in their uncommon energy and in their intelligence. He trusted her, and let her correct, and sometimes draft, his speeches. Some thought her perhaps the sharper of the two, and certainly the shrewder judge of people, though she was inclined, like him, to bully anyone she thought was weak.

The introduction of such a powerful personality would have upset any family; to the tightly knit Stanleys' Velma was as welcome as a tornado. Some of C.O.'s sisters liked to telephone him several times a week, but while they recognised him as the family's leader they bridled at advice or criticism from Velma. She could be generous – she gave all C.O.'s nieces moleskin coats when they turned 21 – and her stylishness fascinated the younger members of the family. But she had little sympathy for the Stanleys' Irishness, and was furious when Eddie took his daughters to see their grandparents' old house in Cappoquin. She thought humble roots should be cut off and forgotten.

Money was a frequent cause of upset. Apart from Pearl, a businesswoman in her own right, and Ginger, whose husband Donald Beer ran his own Pye-connected company, none of the siblings was well off. Eddie, though a director of Pye and other of C.O.'s companies, was chronically overdrawn. His wife Stella was as good-natured as he was, and their relations with C.O. and Velma were usually warm even if his family thought C.O. enjoyed seeing his elder brother come to him for money. When Eddie took Stella on business trips abroad Velma checked his accounts to make sure they did not use the grand hotels where she and C.O. stayed.

By the end of the 1950s C.O. had set up eight discretionary trusts with total assets of £1 million. Tax avoidance was certainly one purpose; C.O. explained that the earlier trusts dated from the time 'when one … got rid of any funds one ever had in case the Socialists got their hands on them'. In fact he never quite got rid of the money for he dominated the

Figure 8.3   C.O. Stanley and his wife, Velma, 1960

trusts' management and was prepared to use them as almost personal assets, on one occasion suggesting that a newly created trust keep its funds with a bank where he had a large overdraft because it would improve his credit there. The trusts did have the genuinely charitable purpose of helping needy Pye employees as well as giving money to good causes, but they were also designed to help family members from whom he preferred to keep a distance. Family disputes distressed him,

Figure 8.4   Donald Beer (Ginger's husband) with his Pye 'Baby Q', 1937

though he could not always avoid them. Rue's teenage daughter Sally resented what she felt was the patronising of her family by C.O. and Velma and refused to accept the payment of £3 a week (a worthwhile sum in those days) that one of the trusts made to all C.O.'s nephews and nieces when they reached 18. Summoned to Lisselan to explain herself she found C.O. sitting in the library, but he was so uncomfortable at

having to confront a rebellious girl that he held the newspaper he was
reading in front of his face throughout the interview.

This awkwardness in personal contacts promised no good for relations with his own son and in 1953 they entered a new and momentarily
difficult phase when John announced his engagement. 'John has
decided to get himself married', C.O. wrote with no hint of enthusiasm to a business friend he was planning to see in Australia. 'I have
to stay in England until that function is over.' But neither he nor Velma
attended the wedding, which was held in a registry office because John's
bride, Elizabeth Haden-Guest, was Roman Catholic. C.O. had been
brought up to believe that the Church of Ireland was the foundation of
the Stanleys' identity. Belonging to it marked the family out from the
Catholic world around them, and allowed C.O. to think of himself as a
West Briton, someone who was both English and Irish. To appease his
father John undertook to bring up his children as Protestants.

In Ireland, and among the Irish, C.O. was determinedly unsectarian.
He would say to Irish friends, 'I'm just a Proddy-woddy', which they
took to mean he saw it as his duty to integrate with Catholics. This did
not stop him protesting when the Catholic church's domination of Irish
life affected his own interests. In 1953 the newly appointed Bishop
of Cork and Ross asked large local firms to give £1000 each to build
five new churches. As a director of the textile firm Sunbeam Wolsey,
Cork's biggest employer, C.O. found this objectionable. 'Whatever the
religion I am opposed to this sort of thing', he told Sunbeam's managing
director Declan Dwyer, adding he had recently said the same thing to
the Archbishop of Canterbury in England. If it were true, as Dwyer told
him, that 95 per cent of Sunbeam's workers were Catholic, what about
the five per cent who were not? Shouldn't the firm give a proportionate
donation to their churches too? 'I am not being difficult', he insisted.
'I am being honest.'

In truth he was being both, and later directed the same qualities
against his own church when approached to help Bishop Foy's, the
Church of Ireland school in Waterford that he was so proud to have
attended. In 1962 the school's governors asked him to contribute
£10 000 to renovate its dilapidated buildings as a first step to attracting
the pupils it needed to survive. For the next three years, and in spite of
the growing demands of Pye's worldwide affairs, he took the time to
lecture bishops and other Church of Ireland worthies on the need to run
the school in an efficient and what he called 'liberal' manner, eventually setting up an £80 000 trust fund to build the entirely new school he
thought was needed. His condition was that its management be brought
up to date, and he lost patience when the Church proposed installing
a board of governors that still included the Bishop of Waterford as
chairman, three members chosen by the Anglican diocesan council,
and one by the heirs of the school's founder. These were just the sort

Figure 8.5   John Stanley and his wife, Elizabeth, after their wedding, 1953

of 'feudal practices' he wanted done away with, he scolded the governors. Why should a Bishop of Waterford sit in perpetuity over the governing body? Why should Anglican clergymen continue to dominate Bishop Foy's management when the school's 'record over the past forty years and its present sorry state is a very poor recommendation for continuing this type of [clerical] control?' He withdrew his offer, and shed no tears when his old school disappeared, explaining to his friend Alan Bradshaw that it had 'refused to have any liberal point of view and insisted on keeping governors who were ' "dyed in the wool" hunting, shooting and fishing'.

C.O.'s unease about John's marriage was gradually diminished by the tact with which John's wife treated her parents-in-law. It also helped that Liz produced four grandsons, not least because it eased Velma's frustration over her own childlessness (when a none too well-off Rue was expecting her third child Velma had proposed that she and C.O. should adopt the baby). With the new generation in mind he bought three old coastguard cottages on the shore of a little bay called Mill Cove a half hour's drive from Lisselan. Reached by farm track, it was almost as isolated as the Cunnigar where he had spent the summers of his own childhood. The comfort, though, was considerably greater thanks to Velma, who had the cottages made over, built a swimming pool, and sent servants and gardeners from Lisselan to prepare the house and plant flowers at the start of each holiday season.

The grandparents' visits to Mill Cove were occasions for excitement and apprehension. Velma could be good with the children. She brought unusual presents from abroad and delighted the children by her readiness for mischief. When two of John's sons caught sight of her in her bath at Lowndes Place she told them to bring up stools, sit down and watch. C.O. was more distant. He taught Rue's children to sail on their holidays at Mill Cove, making them put toothpaste on their noses to protect against sunburn. He also bought the first two Firefly sailing dinghies in Ireland. But he did not know how to talk to the very young, and instead repeated favourite aphorisms such as 'never forget education is the most valuable thing, boy.' What the children did like was his love of gadgets: the automatic log splitter from Canada, the latest American outboard motors, and plastic lobster pots that he imported from Norway ten years before they caught on anywhere else.

For John and Liz the grandparents' visits were uncomfortably close to a school inspection. Velma, who always drove, signalled their arrival with a hoot from the horn of their approaching car. C.O. greeted John with a 'morning, boy'; Velma called him 'Johnny darling'. The custom was that C.O. and Velma brought their own luxury picnic with them, and ate it at the edge of the swimming pool, an irresistible distraction to the children who were eating with their parents at a table on the terrace. After lunch Velma, dressed as for Belgravia and smoking cigarettes through a black holder, inspected the house. She had no inhibitions about criticising Liz in front of other people, and having pointed out the faults in her housekeeping arrangements and ordered the necessary corrections retired to the car for a nap. C.O. escaped to the water, where Velma never ventured, taking a dinghy to a rock in the mouth of the bay from where he could dive into the sea, or rowing far out trailing mackerel lines behind him.

An Irish journalist once asked C.O. if he had invested in Ireland for patriotic, or purely business, reasons. 'From pure bloody sentiment', he said. 'I was well aware that investments outside Ireland would have been more profitable.' His love of the country was always mixed with exasperation. When Rue was looking after their mother at Lisselan before the war she wrote C.O. a letter that conveyed the ambivalent feelings of the achieving Stanleys for the impoverished world they had escaped from. She had passed through Cappoquin and told her brother that she found it 'dead'. Sir John Keane of Cappoquin House

> enquired how the ✶✶✶✶✶ you made your money, and said he heard you were paying Super Tax ... The inhabitants have certainly gone to pot ... Aunt Jane that coarse old woman of eighty three asked me where my rich brother was that he was forgetting she was a poor relation. I am afraid I told her that she was NO relation and that you worked hard for what you had and that it was up to everyone else to do the same. She said she heard you used pound notes instead of Bronco, only much more coarsely put, the wicked old thing.

C.O.'s first Irish venture, a three-man branch of Arks that he opened in Dublin in 1930, led the following year to his involvement in Sunbeam, the Cork textile firm that was unusual for being owned by the well-to-do Catholic William Dwyer at a time when Protestants still dominated business in the city. According to Rue, Dwyer had asked C.O. to do some advertising for him and C.O., perhaps having picked up gossip in the Cork and County Club and suspecting a hidden motive, sent Rue to Sunbeam to make discreet enquiries. After discussing the proposed publicity work with Dwyer she learned from his secretary that he had only just managed to pay the last week's wages. C.O. gave Dwyer £5000, later added another £10 000, and became a director. Renamed Sunbeam Wolsey thanks to an arrangement with C.O.'s well-known British advertising client Wolsey Knitwear, it was by 1966 Ireland's sixth largest industrial company and, with 3500 workers, its biggest employer.

C.O.'s relationship with Sunbeam was not always easy, and two years after helping save the firm he complained he had been 'blackguarded ... for weeks' by William Dwyer and was tempted to 'send him to hell'. The quarrel, cause unknown, subsided and Dwyer came to value his advice on financial matters. Later he won C.O.'s admiration for ignoring nationalist disapproval and choosing to work in London for the wartime Ministry of Supply. Dwyer was an eminent Catholic layman, praised by the clergy for having built a church outside Sunbeam's gates that local wits called 'Willie Dwyer's fire escape'. The partnership suggests how determined C.O. was to conduct himself as 'just a Proddy-woddy' in a Catholic world, and as he expanded his activities in Ireland he never lost sight of Irish sensibilities.

Figure 8.6   The Sunbeam Wolsey factory at Millfield, Ireland, 1942

Nor did he lose sight of the opportunities for a businessman in a country trying to escape from economic backwardness. He set up Pye (Ireland) in 1936 to exploit a new Irish tariff on imported manufactured goods, ensuring the company's own Irish credentials by putting William Dwyer on the board and appointing Pye's Dublin agent J.P. Digby as managing director. This allowed him to escape the import duties his British competitors had to pay, an advantage Pye (Ireland) rubbed home with the advertising slogan, 'Duty Free Radio'. Given the limitations of the Irish market, Pye's Dublin factory was never large, and by the 1960s had only a tenth of the workforce employed by Sunbeam. The company was always dependent on Cambridge in technical matters, but remained moderately profitable until the Irish government began to phase in free trade in preparation for joining the Common Market. Pye (Ireland) was important to C.O. because it established his credentials as an Irish manufacturer rather than an Irish-born entrepreneur who had made a fortune in England, but the best proof of his commitment to Ireland was the Aberdare Electric Company, later and better known as Unidare.

A columnist for the *Irish Times* wrote that 1947 would be remembered a century later as the year when

> somebody switched on some lights in a village ... and rural electrifi-
> cation took its bow [in Ireland]. And if that does not mean more to the
> country than the rest of the year's events put together I shall be very
> surprised indeed.

Life in rural Ireland had changed little since the start of the century, not least because of the lack of electric power, and politicians feared the countryside was so set in its ancient ways they thought it necessary to ask parish priests to explain the benefits of electrification to their villages. When the government built a hydroelectric plant on the Shannon River it had no choice but to import all the equipment from abroad, causing havoc with Ireland's balance of payments. This gave C.O. the idea of setting up an Irish company to make at least part of the equipment the Irish Electricity Supply Board (ESB) needed to bring electricity to rural areas. The result in 1947 was Aberdare Electric, with a factory at Finglas to make transformers and overhead conductors. It was a brave gesture, for at that time it was hard to recruit the necessary scientists and engineers in Ireland, and Irish opinion doubted a local company was capable of making heavy industrial equipment. At first the ESB was Aberdare's only client and might have remained so, C.O. admitted later, so 'strong was the prejudice of any potential customer [against] buying Irish'. He sometimes himself complained about the Irish way of working. Told that Unidare could manufacture a new polythene field drain he commented that it was 'a lazy way of doing ... draining and [therefore] might appeal to the Irish'. He thought Irish farmers would

Figure 8.7   Advertisement by C.O. Stanley's publicity company, Arks, for radios produced by Pye (Ireland) Ltd., 1937

Figure 8.8   C.O. Stanley with his sister, Pearl, Alan Bradshaw (beside Pearl) and Charles Harmer (on the other side of C.O. Stanley), early 1950s

never work as hard as the Dutch, though having observed the latter's dour way of life wondered whether the Irish were not right to take time off for enjoyment. But he never doubted the quality of an Irish industrial workforce provided it was well led and 'educated to get rid of the shadow of the gunman'.

C.O. became Aberdare's chairman and its directors included his old Irish associates Dwyer and Digby, and also Alan Bradshaw and Robert Renwick. Bradshaw had gone out to Ireland as C.O.'s eyes and ears soon after his return to Pye from wartime duty at Bletchley Park. He was the only member of Pye's senior staff who socialised with C.O. and Velma as an equal; the two men played golf and the Bradshaws' visits to Lisselan were occasions for merrymaking. Canny, straightforward and loyal, he was perhaps the only real friend C.O. had inside the growing Pye empire.

Unidare was quickly successful, expanding far beyond its early contracts for the Irish government, and by the 1960s was Ireland's seventh largest employer with a workforce of 2500. C.O. made sure that Irish nationals owned a good proportion of its stock, and at one time both the Roman Catholic Archbishop of Tuam and the Protestant Archbishop of Dublin were among shareholders. The precaution stood him in good stead when the company's success tempted foreign predators. To avert the danger the Irish government allowed Pye to acquire 75 per cent of Unidare's shares, and sell on a third to the American aluminium giant Alcan with which Unidare was already associated. This was judged to be 'consistent with the interests of Eire and the Irish minority share-holders' who held the remaining 25 per cent of the stock. Bradshaw told C.O. that the Commerce Ministry only agreed to this arrange-ment, which some politicians had attacked for letting Unidare slip into 'foreign hands', because of 'your standing and that of your Irish asso-ciates' (Unidare's Irish directors now included the former head of the ESB, R.F. Browne). C.O. still felt it necessary to pledge in public that, should Pye itself be taken over, it would offer Irish citizens the chance to buy a majority holding in Unidare – a promise he would not be able to keep.

People listened to him, even held him in awe, but also feared him for his abrasive tongue. And not everyone accepted his Irish creden-tials. The local business leaders, known as the Princes of Cork, never fully accepted someone they considered not only cranky but also half-British. It was different at Lisselan. There he was 'the Boss', powerful and demanding but also benevolent. He eventually acquired more than 900 acres of land. It was of poor quality, but with Aberdeen Angus, Herefords and Shorthorns he aimed to make a model farm to point the way towards a modern Irish agriculture. Knowledgeable British visi-tors might think the farm 'mucky', but C.O. had a milking parlour as good as any in Britain and won prizes with his pedigree cattle at the Royal Dublin Show. He designed his own farm buildings and had the first silage pit and the first combine harvester in County Cork.

Most mornings at Lisselan began with a visit to the farmyard to talk to the men before they set off to work. In a rural Ireland where the government still provided few services he won hearts by practising, but with a difference, the paternalism that was expected of landowners such as he. Tradition demanded that estate workers got Christmas boxes of money, cigarettes and a joint of meat, but Lisselan's employees were also given cottages that had electricity, kitchens with a modern stove and clothes drying cupboard, and a bathroom with hot running water and flush water closets (the latter attracted admiring visitors from all over the county). Velma organised a station wagon to take the estate families' children to the village school, washed the hair of the children of problem families, and sent food to the neighbourhood sick regardless of whether they worked for the Stanleys.

Velma might seem a brisk Lady Bountiful, but C.O. liked drinking and talking in the local pubs. When country life came to a stop on Sundays and the frequent Irish church holidays, a practice he hated, he drove members of the staff to mass and back. He won hearts by fairness and discretion; people said 'you could always go to him if you were in trouble. He'd help, and never tell a soul about it.' Above all he was ready to support anyone who showed a desire for education, paying for bright Lisselan children to go to secondary schools in nearby towns, and later helping find the right college or job for them.

Education had saved his own family from an uncertain future, and he believed it was the key to pulling into the modern world a country dominated by a conservative church and, as he put it, too long 'interested in politics and bored by economics'. In the days of British rule most industry had been concentrated round Belfast and so was lost to independent Ireland. In 1932 he had been asked to give advice on industrialisation to the future prime minister Eamon de Valera. His message was that Irish industry had to be based on a healthy agriculture, and that there was little value in Ireland assembling radios, as he was then proposing to do, as long as it did not have an industry to process its food and make goods for its farmers. By the 1950s, with his own experience at Lisselan in mind, he was talking provocatively of the 'drudgery and hardship of rural life' in a country still little changed by the 20th century's agricultural revolution. He compared Ireland to a similarly agricultural society such as New Zealand, and found the difference shocking.

He knew, and did not care, that his outspokenness laid him open to attack. 'With the sensitive political system we have here . . . if you dare to criticise a politician you will be termed a West Briton, a Cromwellian or a Communist.' This was why 'few people have the courage to be honest', though on the subject of Ireland's backwardness he could quote in support an unlikely ally, the conservative Bishop of Cork, Dr Lucey. The bishop had declared that a 'true Irish patriot' did not fix his gaze on the lost six counties of Ulster but on the 'eight hundred thousand men and women lost to the nation since 1922' by emigration. In a speech in Cork in 1957 C.O. asked what point there was in saying, as de Valera had just done, that Ireland was a wonderful country 'when few of those that leave it ever return'.

At times the Irish press treated him like a hero, and an Irish hero at that. When he fought the British telephone manufacturers' cartel in 1960 the *Irish Times* chose to describe him as a 'Clonakilty farmer', and celebrated the 'Irish David who looks like a local schoolmaster' and had dared stand up to powerful British monopolists. But when the Clonakilty council and local businessmen asked him to bring industry to the area he told them nothing would change as long as Irish farms were too small to be economic. And it was on the issue of farming that he became bitterly engaged with traditional Ireland in the formidable person of Bishop Lucey.

At a confirmation ceremony in the spring of 1967 Lucey made a reference to a 900 acre estate 'built up by a non-resident industrial magnate through the purchase of farm after farm . . . Nine hundred acres of good West Cork land [was] rather too much for any one individual to accumulate as a sideline', particularly if it was the result of buying up

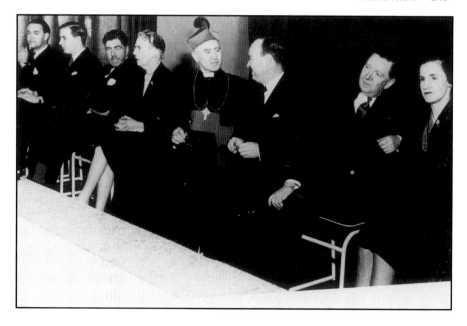

Figure 8.9   C.O. Stanley in company with Bishop Lucey for the opening of a Sunbeam Wolsey factory in the 1960s

small homesteads. The Bishop alluded to a 'rumour' that half the estate was now to be sold off quietly 'and not to a Catholic'. This reminded Dr Lucey of the 'secret sale and no papist rule' that applied to land sales in the days of the British, and he called for the Irish Land Commission to acquire the acres in question by compulsory purchase and offer it to landless farmers and workers on the Lisselan estate.

A more prudent person might have kept quiet until the storm blew over, but C.O. stirred up the controversy by letting Velma send a fiery rebuttal to the *Cork Examiner*. She reminded Dr Lucey that the Lisselan estate had improved land that was once derelict and overgrown, while her own work 'over thirty years has been to turn hovels into homes'. When C.O. paid to send estate workers' children to school, 'the fact that they were all "papists" (to use the Bishop's term) was neither here nor there to us. They were Irish, with very little chance of other than the village school', who could now perhaps become 'good liberal Irish citizens'. As to selling the land to Protestants, 'we did not ask our agents if the [would-be] purchasers were Catholics or Protestants'. Lucey, she concluded, should take care his words did not 'set a match to actions which most Irish people would deplore'.

The controversy was too public for the Land Commission to ignore, and it issued a compulsory purchase order for the several hundred acres C.O. had already agreed to sell privately (the estate had always lost

money, and he now wanted to cut expenses). It made no difference finan-
cially whether he was paid for the land by the Commission or the two
original would-be purchasers, but it had become a matter of principle
for him and he took the dispute to Sean Lemass, de Valera's successor
as Taoiseach and admired by C.O. for his determination to modernise
the Irish economy. 'Great principles' were involved, he told Lemass,
and the affair was 'one of the most disgraceful things I have encoun-
tered'. Not only was the Commission proposing to sell off the land
in smallholdings that would not be economically viable, but Lucey's
blast meant that the affair could only damage 'our image as "Tolerant
People"'.

The compulsory purchase was for C.O. one more example of Irish
politics getting the upper hand over economic sense. Of course at times
he played the political game himself. Told by the company secretary of
Sunbeam Wolsey that Fianna Fail, in William Dwyer's opinion the 'best
of a bad bunch' among the Irish parties, was likely to ask for a campaign
contribution, C.O. pointed out that some firms made a large payment
to its preferred party and a smaller one to its opponent. He took care to
add that 'this is only for your information and not a recommendation',
though this may well have been his own practice at Pye (Ireland) and
Unidare. It was dubious territory for a West Briton, and he was in any
case sceptical of the usefulness of such payments, a scepticism that was
reinforced by his attempt to bring television to Ireland.

It began in 1950 when the BBC was considering starting its television
service in Ulster. C.O. warned that British broadcasts in the North would
reach parts of the Republic and that 'this would create a great deal of
discontent from the people who want to see television from their own
Irish station'. He was right, and as the 1950s wore on politicians in
Dublin were increasingly embarrassed by the growing number of Irish
citizens who bought sets to watch Ulster's television. In 1951 C.O.
put on Ireland's first demonstration of television at the Royal Dublin
Society's Spring Show, and tried to appeal to Irish sensibilities. He
called television 'a great art' that would allow Ireland to 'continue the
tradition of our Dublin theatre'. And he suggested television could
reduce emigration by bringing entertainment and education to the most
abandoned parts of the countryside.

The politicians said Ireland could not afford a television service, but
C.O. claimed he could build a transmitter for the Dublin area for no more
than £50 000. 'Now the fat is in the fire', thought the *Irish Times*, for if
C.O.'s figures were correct ('and he is a person of the highest eminence
in the radio and television world') there was no reason why the country could not have two or even three such modest stations. The *Times*
was too optimistic, and as years passed and Ireland still did not have its
own television, C.O.'s demonstrations of impatient energy and superior
expertise began to do him more harm than good. He hired the former

Minister for Posts and Telegraphs, Erskine Childers, to help his consortium of Pye (Ireland), ATV and the American broadcasting giant CBS to draft proposals for an Irish television service that met government concerns. Before long, though, Childers' over-eager lobbying on Pye's behalf alienated important members of the Television Commission that had been set up to decide the country's television future. C.O. muddled matters further by challenging the government's choice of site for the first transmitter, arousing suspicion that he was looking for a cheap and inferior technical solution. Nevertheless the Pye consortium remained a front-runner for the contract, not least because C.O. had worked hard to give it Irish credentials. But Irish officialdom was a constant source of exasperation, and he was less surprised than most when in 1959 the government performed an about-turn and gave the new television service to a statutory authority rather than a commercial company. Writing to Norman Collins a few years earlier he had confided, 'I hate people assuming that we are going to get the [television] concession [here]. I never believe anything in Ireland until it happens.'

Ireland did not always disappoint him. Passing through Cappoquin in 1960 he had a chance encounter with Michael Sargent, the town's motor dealer and garage owner who, as a boy, had been taught to row by C.O.'s father. Sargent told him about the decline of the Cappoquin Rowing Club. Many of the old Irish clubs had disappeared and Cappoquin's was struggling for money and seemed likely to follow them. In spite of his reverence for his parents C.O. usually kept clear of his birthplace, not least because he did not want to be drawn into small-town feuds and jealousies. But he liked Sargent, who had a persuasive tongue and a store of memories of shared boyhood on the river that he called 'the old Irish Rhine'. He also flattered C.O. by comparing him to his father – 'the same gestures with the fingers, the same laying down the law'.

Sargent did not know that C.O. was already thinking of buying H.C. Banham, the Cambridge firm that built racing boats for the university and college crews. Banham joined the growing number of Pye subsidiaries, and the two boats C.O. had made there for Cappoquin allowed the Club to celebrate its 1962 centenary in style. Sargent described to C.O. how the two new craft, a Fine 4 and a Fine 8, were laid out on the Rowing Club dance floor, where 'the whole of the community big and small [and] including many of your Dad's pupils' inspected them. The centenary celebrations began with a blessing of the new boats, which Velma then christened. The Keanes gave a lunch at Cappoquin House and C.O. became the Club's honorary life-president.

Pleased though he was to honour his father's memory, C.O. did not like to be seen as a public benefactor, preferring to be recognised for his achievements rather than for his generosity. Towards the end of the 1950s Francis Coulter, a friend since the early days of Arks, suggested

*Figure 8.10   C.O. Stanley in procession at Trinity College, Dublin, when he received
an honorary Doctorate of Laws, 1960*

that Trinity College, Dublin, where Coulter had studied, should award
C.O. an honorary degree. Trinity at that time did not see businessmen as
suitable candidates for honours, and Coulter had to open the Provost's
eyes to his friend's achievements which, he told C.O., '[were] news
to him'. The Provost in turn warned Coulter that some conservative
members of the College Board thought 'accomplishment in the industrial field ... of less consequence than some bucolic parson's elevation
to a meagre diocese'. The Provost invited C.O. to a lunch which did
not go well, for he left under the impression that any honour Trinity
might give depended on his making a donation – exactly the sort of
deal he would not accept. He was happy to give money, he explained
to Coulter, but 'I certainly would not take any recognition from anybody because I had done something I thought was worth doing on its
own account.' In fact the Provost had noticed that his lunch guest was
'rather sensitive on the question of honorary degrees', and made it plain
that Trinity was attaching no strings to the honorary Doctorate of Laws
it offered him. C.O. received it at a ceremony in 1960 that also honoured Oxford's Irish-born French scholar Enid Starkie and the left-wing
British publisher Victor Gollancz; C.O. was the only businessman on
the list.

The citation for his degree began,

Inter miraculi aevi nostri ... Among the wonders of our age so lavish in wonders nothing is more wonderful than the instruments that enable sounds and images to be sent and received from a distance by means of electrical waves ... Among the first rank [of his industry C.O. Stanley has given] timely help to the Irish government in their industrialisation policy ... His energy, foresight, skill and public spirit [has been] demonstrated in peace and war; in the Western world and in the antipodes.

The archaic phrases gave dignity and context to achievements he had often won against the grain of contemporary opinion. However contradictory his feelings towards Ireland, there was nothing ambivalent about his pride in Trinity's honour, and he wore its tie almost every day for the rest of his life.

Sources

COS files: 1/5, 2/1, 2/12 (Pye Ireland), 2/13, 2/20, 3/4/1–3, 5/1, 6/1 (Trinity College, Dublin), 7/1/5.
COS/VDS1/5 (land dispute), COS/TWEM2/2–3.
Unidare folders (including company reports).
Simoco boxes: Box 1 (Cooper Bros report on Pye /Ireland). Box 3 (Pye (Ireland)), Box 4 (Irish TV).
Pye main board minutes.
Interviews: Brian Beer, Sally Emerson, Rethna Flaxman, Nuala Hall, Fred Keys, John O'Connell, Nicholas Stanley, David Stewart, Warren Tayler, Daphne Whitmore, Michael Worsley.
Michael Bell interviews: Dillon Digby, Tom Linnane.
A–Z files: Dillon Digby (Pye (Ireland) annual general meetings). Mrs D. Dwyer (Sunbeam Wolsey papers).
Books: Savage, *Irish Television*.

Chapter 9

Danger years

In 1959 Robert Browning, a Colonial Service officer whose posting to Cyprus as private secretary to the governor had just ended with the granting of the island's independence, took on the job of C.O.'s personal assistant. Eyes sharpened to his new British surroundings by years spent abroad, Browning found a good deal to puzzle him at Pye. On his trips to the United States C.O. had picked up the American enthusiasm for open-plan offices, and the long top floor of the new Pye headquarters on Chesterton Road was little more than a glass-roofed shoe-box. Only C.O., with an office at one end, and the directors, with compartments along the shoe-box's sides, had their own private space.

Browning observed that this supposedly democratic layout had little effect on those who worked there: life in the shoe-box was characterised by strict, if informal, hierarchy. The female staff formed a pyramid, at its apex the Secretaries (always spelt with a capital S), and the girls who collected and delivered the office mail at the bottom. The latter called the Secretaries 'toffee-nosed', the Secretaries thought the post girls 'beneath contempt'. Men were less rigidly divided, but directors with their separate dining room (a privilege B.J. Edwards had insisted on) were naturally at the top, while the lowliest position belonged to a kindly complaints clerk who dealt with dissatisfied customers and, to Browning's mind, was the most valuable member of the office staff.

The one person untouched by these jealous distinctions was C.O., who seemed unaware of the battles that raged beneath him. Nor did he see any contradiction between the modern spirit of his head office's design and its old-fashioned wage system. Everyone, from directors down, received small salaries, often unchanged since the war, that C.O. topped up with bonuses he awarded at the end of each financial year. The result, Browning observed, was that no one could afford to put a foot wrong because their own and their family's well-being depended on C.O.'s favour. He still inspired affection, and people's respect for him grew with the company's success, but this degree of dependence on one man also bred fear.

C.O. certainly inspired fear in the 24-year-old solicitor Michael Rose, deputed to receive him when his own lawyer Frank Levinson was ill. C.O. was at the time in dispute with his fellow electronics magnate Jules Thorn, who had just sold his shares in British Relay Wireless, a company that provided wired radio and television to hotels and apartment blocks. British Relay's chairman was Robert Renwick, and Pye and Murphy were its chief shareholders and supplied most of its equipment. C.O. accused Thorn of selling his shares without first offering them to Pye and Murphy, as he claimed Thorn had pledged to do. Displeased at not finding Levinson, C.O. flung Rose an angry challenge. 'Before I say anything I want to know if I'll get Thorn in the box.' Taken aback by this choleric little man with 'huge florid cheeks and a lick of black hair', Rose asked C.O. to tell him precisely what had been agreed with Thorn. C.O. said he had 'made it perfectly clear', a favourite expression that in this case and many others ignored the absence of any document obliging Thorn to do anything. This was not an oversight on C.O.'s part. John Stanley explained that his father liked 'to keep agreements and arrangements . . . in a rather flexible situation so that if these went wrong he could claim that was not what he really intended'. Cecil Rieck, C.O.'s private financial adviser since before the war, spent much of his time resolving misunderstandings caused by unrecorded deals.

Even people who did not depend on C.O. for their well-being could feel oppressed by him. It was, after all, Ellis Birk, his colleague on the board of ATV, who noted how he could be 'bullying and tiresome and relentless with people who didn't share his vision'. This was C.O. at the peak of his achievements and his powers, a man who, according to Sir Richard Powell, director general of the Institute of Directors, 'didn't give a damn for anybody . . . [and] believed a rather dangerous thing . . . that God had given him the gift of speech in order that he could say exactly what he . . . [thought]'. The truth was that after Pearl's death from breast cancer in 1956 there was no one left in Pye who dared challenge him. The obituary of Pearl that appeared in the company's annual report did not mention that she was his sister, and gave no idea of her importance to him as an adviser who was never afraid to question his judgement or remind him that his fatal flaw was over-confidence. With Pearl gone C.O. ruled over a docile board. Eddie always avoided confrontation with his brother. Charles Harmer spoke out most at board meetings, but he and the other older directors had always gone in awe of C.O., while the younger ones were all his own creation. From time to time colleagues advised him to strengthen the board with an outsider, but independent nonexecutive

Figure 9.1   Cecil Rieck, c.1949

directors were then rare in British companies and it was easy for
C.O. to dismiss the suggestion. His 1958 appointment of Sir Ben
Barnett, former senior civil servant at the Post Office, to the boards
of both Pye and Telecomm brought excellent contacts in those parts
of Whitehall that most interested C.O. He said Barnett had a 'wonderful brain', but as a director he proved no more eager for confrontation
than Eddie (a Pye colleague thought Barnett a 'delightful man who
couldn't run a sweet shop'). On the one occasion he did make an
uncomfortable suggestion at a board meeting C.O. brushed it aside.
Anyone who knew how he ran Pye would have been amazed had he
done anything else.

Five years before bringing in Barnett, C.O. had expounded his ideas on directors and management in a letter to a business acquaintance:

> I believe in running a business with the help of men who are objective and never "yes men". I hope that every one of them has the ambition to pinch my job, and it is of little importance to me whether they have degrees or from what families or schools they have come.

It was true he never worried about a person's educational qualifications or social background, but the rest of what he wrote was fantasy, and raises the possibility that he did not understand the nature of the world he had created around himself. Michael Nathan of Pye's auditors Howard Howes thought C.O.'s board was 'cowed'. After observing C.O. at close quarters for almost three years Robert Browning decided he had surrounded himself with 'yes men', and wondered if this pointed to a hidden sense of insecurity in his employer. The subservience of the Pye board certainly sat uncomfortably with the company's continuing expansion and the proliferation of the subsidiaries of which C.O. was so fond. The diversity of Pye's operations astonished Browning. They now ranged far beyond its traditional manufacture of television sets and radios and early post-war ventures such as Telecomm.

A series of companies was developing the instrument-making tradition started by W.G. Pye, and now produced equipment for the nuclear industry and a wide range of industrial work. The skills of Dennis Fuller and Don Weighton allowed Pye to make the world's first electronic pacemaker, the pulse of which could be reset by a radio signal. Doctors liked it and C.O., Fuller remembered, 'was as pleased as a small child with a new toy', but it was overtaken by American firms using space-age components Britain did not have. Other Pye companies dealt in navigation aids and military electronics; while the success of Unidare in Ireland inspired him to set up a company to make transformers and heavy electrical equipment in Britain. C.O. had gone into gramophone records and after an erratic start beat EMI to make the first British stereo recordings, though the company only began to make profits when his show-business friends at ATV acquired a half-share and management rights in 1958. Pye companies made or distributed consumer goods such as electric heaters, toasters, and the 'Hostess' heated trolley for carrying food from kitchen to dining room. Pye directors did not think it odd when a board meeting was interrupted to allow the president of the American Ironrite Corporation to demonstrate his latest patent iron.

C.O. also expanded his nonbusiness activities by fund-raising for the Cambridge Arts Theatre, which nurtured several great British stage careers but was often strapped for cash. In 1959 he threw the first of several dinners for Cambridge businessmen to persuade them to contribute to the theatre, and at the end of his speech took his own cheque from his pocket and waved it the air. Dadie Rylands, the King's

Figure 9.2   Pye broadcast equipment undergoes vigorous testing, 1952

Figure 9.3   Raising wreckage of the Comet, located by Pye underwater television cameras, 1954

Figure 9.4   Front cover of a leaflet, 1955

don who was the inspiration of the Arts, thought it 'great theatre'. It was the culmination of an unusual friendship that began in the war when C.O. made time to go to Rylands with money and advice after Maynard Keynes, the great economist who guided the theatre's finances, went to work in the Treasury. C.O. got on so well with the intellectual Rylands

Figure 9.5   Manipulators made by Pye for handling 'radio-active' material in the Bond film 'Dr No', 1962

that they began to meet regularly for lunch, and in 1962 the Arts Theatre made C.O. a director and trustee.

The result of Pye's expansion was that by 1965 it comprised 109 companies of which 39 were based abroad, the most important of them in Australia and New Zealand. C.O. had built these up on frequent post-war visits when he formed a close friendship with Arthur Warner who, like C.O., dominated a family electronics business. Pye's accountants usually found out about new acquisitions only when their team arrived in Cambridge to conduct the annual audit. 'At other companies we would be asked to investigate and give our opinion on possible subsidiaries', commented Howard Howes' Walter Meigh. 'I don't think any other client [of ours] went into businesses without investigating them as Pye did.'

C.O.'s 1960 takeover of Temco, the Telephone Manufacturing Company, showed how he set about an acquisition. At that time the manufacture and pricing of Britain's telephone equipment was controlled by TEMA, the Telecommunication Engineering and Manufacturing Association, a ring of companies that included AEI and GEC. Telephones were an obvious area of interest to Pye Telecomm, but when C.O.

Figure 9.6   Nuclear reactor with revolutionary transistorized control gear, built by Pye in collaboration with the UK Atomic Energy Authority, 1964

applied to TEMA he was told there were 'more appropriate associations' for him to join. Rebuffed by a monopoly, and a backward one at that – in 1957 Britain had 13.46 telephones for every hundred inhabitants compared to 33.73 per hundred in the United States – C.O. resolved to break his way in. Troubles at the TEMA-member Temco gave him the chance to make a takeover bid. Seven TEMA members contested the Pye offer but were outmanoeuvred by C.O., who at one point in the battle telephoned AEI's chairman Lord Chandos pretending to be an Irish newspaper reporter to ask how TEMA's counterbid was doing.

C.O. got Temco, a victory that made him a hero in unlikely places. Even the liberal *Observer*, which had fiercely opposed him over commercial television, praised him as the 'cold-eyed but amiable Irishman . . . whose personality makes Pye the most newsworthy firm in a newsworthy industry'. In fact the takeover was a less than brilliant stroke, for Temco was not the business C.O. thought it was. Pye paid £2.5 million in the belief that it was buying a company that made telephones; what it got was a company that only assembled them. C.O. had deputed Keys, Harmer and the new managing director of Telecomm, John Brinkley, to supervise the acquisition. In the Pye tradition they carried out only a superficial audit and when the truth about the new

Figure 9.7   *C.O. Stanley and Sir Robert Renwick (standing) at a fund raising lunch in Cambridge, 1960*

subsidiary became known Brinkley shredded all the paperwork connected with the takeover. In the words of a senior executive, 'Pye was sold a pup'.

C.O.'s hasty acquisitions were tacked onto a group where overall control was already shaky. Gordon Maclagan never saw a proper central or group budget in all the years he worked as managing director of the Pye subsidiary Newmarket Transistors. This did not surprise him because he had earlier turned down C.O.'s offer to make him Pye's chief accountant, recognising the appointment as a poisoned chalice. C.O. still went through the motions of looking for a finance director or group accountant, but those he did engage never stayed long. Their departures did not dismay him because he could not accept he needed the support of a first-rate accountant. At the same time his financial factotum Fred Keys went out of his way to make life difficult for anyone C.O. did appoint. Michael Nathan considered Keys a 'model company secretary', but quite out of his depth in Pye's increasingly complex post-war affairs and yet so jealous of his relationship with C.O. that he 'thwarted access [to him] by brighter people desperately needed in the finance department'. Once when C.O. appointed a finance director at Nathan's urging the man resigned after a few months because Keys would not let him attend board meetings. C.O. allowed Keys to get away with such behaviour because he found him useful, not because he had a high opinion of him. Keys bored him, and the annual lunches to which he and his wife were invited at Sainsfoins were a trial to both C.O. and Velma. The relationship so prized by Keys appears on C.O.'s side to have been entirely cynical.

In the late 1950s C.O. set out on a new strategy to reduce Pye's dependence on radio and television sales by stressing the performance of the growing tribe of subsidiaries. The new strategy involved supporting to the maximum already profitable subsidiaries such as Telecomm, building up all other promising companies outside radio and TV, and feeding development capital only to companies that worked in vital future technologies such as transistors and cathode ray tubes. It was this thinking that led him to tell John Brinkley in 1960 to double Telecomm's profits in 5 years. C.O. knew the market for radio and TV was fickle enough to jilt even him, hence his determination to neutralise what he called the 'terrible peaks and dips' that affected sales of receivers. If this had remained Pye's key market for so long it was, he admitted, because 'that is where the cream really comes from'. In the first half of the 1950s Pye got more of that cream than any other manufacturer, winning over 20 per cent, and for short periods as much as 30 per cent, of the British receiver market. The opening of commercial television in 1955 had boosted sales again, and in March that year Pye announced record profits of £2.2 million. These rose to £2.3 million in 1956, almost double the figure of 2 years before.

But the rise in demand was temporary, and falling sales were further depressed when in the course of 1955 the government again reined in the economy. From this moment Whitehall policy caused havoc in the television business. The quickest way to discourage consumer spending was to reintroduce restrictions on hire purchase that had been lifted completely in 1954. They were now progressively tightened until 1956, when the minimum deposit requirement reached 50 per cent of a set's price. In the autumn of 1958, after C.O., Jules Thorn and Ekco's E.K. Cole had lobbied the Chancellor of the Exchequer Derek Heathcote-Amory, all controls on HP were removed, only to be restored in little over a year. Purchase tax, C.O.'s other bugbear, remained in force throughout, though, and by 1959 was costing the radio and television industry £40 million a year, a tenth of all the revenue the tax brought in. C.O. argued that the high level of purchase tax limited the potential ownership of TVs to 8 million. Cut it by half, he said, and ownership would reach 12 million.

The tax was reduced in 1959, contributing to a boom in 1960 which, given the chronic weakness of the British economy, forced the government to scramble once more to reduce spending. This stop–go policy certainly hampered C.O., but it hampered other manufacturers too, and was perhaps less damaging than the faults that had appeared inside Pye

Figure 9.8 Views of Oulton Works, Lowestoft, 1960. Pye opened the factory in 1951 and gradually transferred all television manufacture to it

Figure 9.9   C.O. Stanley (hands clasped) visiting the Lowestoft factory, 1964

itself. Up to 1955 Pye television sets were better than, or at least as good as, those of its competitors. That year Pye produced what C.O. himself admitted was 'a really shocking bad television set'. At one stroke, he said, the company lost both its 'marvellous reputation' and its position as the biggest seller in the market. Pye's pre-tax profits for 1956–1957 fell back to £1.6 million, and would not reach £2 million again for another 3 years. It was, C.O. confessed (though only 3 years after the event), 'the biggest catastrophe that could happen to a business'.

The 'shocking bad set' was the VT 14, which had a tendency to drift and several unreliable components, including valves and capacitors. These faults were evidently the result of both poor design and inadequate quality control which C.O. blamed on too much specialisation. No Pye television engineers lost their job over the 'catastrophe' – he said 'they hadn't deliberately made . . . mistakes' – but he ruled that in future one engineer should be put in charge of the design of each model.

C.O. first mentioned the VT 14's problems in late 1958 at a special meeting of Pye executives to whom he also had to explain why it took him so long to admit the set's failure. He did not say when he first learned of the problem, only that it had taken Pye's TV division most of 1955 to accept the truth of what had happened. It was then that he imposed a policy of silence. 'When you run into difficulties there are many reasons why you do not talk about them.' There could have been panic, with 'good people' leaving the company. More compelling still was the need to keep the 'most valuable asset . . . the Pye company has got', the goodwill of its dealers. He resolved to do this by telling the dealers nothing, and praised the sales department's 'great job' in 'never admitting that things were bad'. He judged it a success that, though dealers eventually gave up trying to sell the VT 14 and its variants, they did not send them back to the Pye factory. Dealers' confidence had nevertheless been shaken, and the company would have to work hard to restore it.

C.O. also admitted that he wanted to keep his troubles hidden from Barclays Bank. During the financial difficulties triggered by the fall in receiver sales Pye's overdraft had ballooned from £112 000 in March 1955 to £4 million 3 years later, though C.O. made it sound nothing out of the ordinary.

> We live on a bank overdraft – who doesn't? – and if this bank . . . got to hear that we had troubles, God knows what they would do. So we had to hide that one up. You must never think that a bank has . . . any great feelings of love for you.

It was not the only thing he tried to 'hide up'. A board meeting in November 1957 heard that that year's radio and TV sales were expected to produce a profit of only £175 000 on a turnover of £13.4 million. To improve those figures C.O. ruled that all television sets in dealers' shops at the end of the financial year were to be recorded as 'firm sales', and not, as was customary, as stock. He instructed Fred Keys to explain this ruse to the auditors Howard Howes and 'deal with any objections [they] might raise long before the end of the financial year'.

C.O. quoted Churchill to support his tactic of defiance. 'Never give in . . . Never yield to the apparently overwhelming might of the enemy'. The trouble was that the enemy was now as much inside Pye as out, as the *Investors Chronicle* hinted in an analysis of C.O.'s acquisition of the transformer manufacturer Lindley Thompson at the beginning of 1959. Pye financed its £655 000 bid by issuing Pye shares to the bankers Guinness, Mahon at a preferential net price of 12s 6d at a moment when their market price was 15s. Shareholders protested at handing what they saw as unnecessary profit to city bankers; the *Investors Chronicle* sympathised with them, but pointed out that a company with a £4 million overdraft, tiny cash reserves and large amounts of stock had a 'very real financing problem'. The paper also published a table comparing the performance of Pye, Thorn, Plessey, Ekco and Ultra in the current decade. Indices of profits, with 1950–1951 as 100, showed Pye easily outstripping the others to reach 372 in 1955–1956, with Thorn a poor second at 236. After 1956 Thorn had taken the lead while Pye went into free-fall, only somewhat slowed in 1957–1958 by the good performance of Tecnico, the Australian valve maker (later renamed Pye

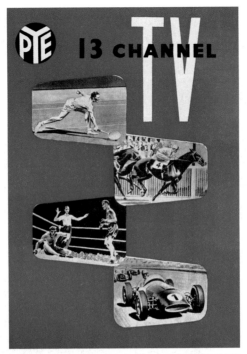

Figure 9.10   Cover of a leaflet, 1955

Figure 9.11 *Prince Philip, Duke of Edinburgh, handing C.O. Stanley an award certificate for the outstanding design of the Pye television receiver, model CS17 at the Design Centre, London, 1957*

Industries) in which C.O. had acquired a controlling interest 2 years earlier.

When C.O. admitted the disaster of the VT 14 to his senior staff he thought the crisis was over. He was frank about the strain it had put him under. After falling ill at the beginning of 1958, he told them, he had gone abroad to recuperate, 'and I never want to have the same feeling when I go away again'. Now the pressure was off. The letters of complaint that he ordered to be brought to his desk each morning had dwindled from 200 a day to a handful and he was in the mood to reinterpret criticisms in the financial press to his own advantage. Tecnico's boost to group profits was proof that his diversification away from radio and TV was paying off. High stock levels were partly explained by the need to build up materials and components for an expanding contract to provide instrumentation for the Spadeadam Rocket Establishment.

What he did not talk about any more were the possible human causes of the sudden decline in his company's fortunes. Perhaps he did not believe it was a problem. Others thought differently. C.O.'s confidential adviser Cecil Rieck worried that Pye's rapid expansion had caused the designers and engineers who produced Pye's successful post-war television sets to move into other fields. Donald Jackson and Leslie Germany were now working on television transmission, cameras and studio equipment, but the biggest blow was the departure to Australia in 1956 of C.O.'s senior television engineer Ted Cope. In the opinion of Cope's colleague Jim Bennett, Pye sets 'never had the same cutting edge again'.

Cope was not replaced, nor, when he died at the age of only 47 in February 1960, was the company's technical prodigy B.J. Edwards. His death was unexpected, though the mounting crisis of his final years was obvious enough. When B.J. Edwards became a full director of Pye in 1951 his reputation for extramarital affairs and hard-drinking was well-established, and Pye's emergence as supplier of television stations to the world provided an opportunity to indulge extravagant tastes at the company's expense. He was ambitious, too. It had riled him that Pearl, a woman, became a director long before he had. Some of his colleagues suspected he hoped to take control of Pye when C.O. retired upstairs to be company president (no one else thought C.O. had any such intention).

B.J.'s behaviour did not improve even after it became plain he was responsible for the 1955 crisis over the ITA transmitter in the Midlands. He was still sometimes offhand with C.O., walking out of board meetings without a word of explanation or apology. He made himself so unpopular with Pye's partners in ATV that Lew Grade refused to attend meetings where he was present. In the spring of 1959 B.J. told the Pye board that he was about to strike a lucrative deal in Cairo in which Pye would receive equity in a company he would not name in exchange for technical information. Pressed for progress reports he prevaricated. The deal came to nothing. He spent large sums of money not only on his increasingly frequent trips abroad but also on cars and a large new house in Cambridge, and at the same time refused to put any money aside for a pension. Colleagues suspected him of taking payment in cash for goods supplied on discount to foreign customers and keeping the money for himself. These damaging stories were eagerly spread by his enemies, among them the equally ambitious John Brinkley. It was even rumoured he had taken a flat in the West End of London where he invited friends to watch blue movies.

Edwards' behaviour was such an open scandal that Peter Threlfall (a future managing director of Pye) could not understand why C.O. did nothing about it and wondered if B.J. was blackmailing him. There is a less melodramatic explanation. C.O. believed anyone he brought into Pye was by that fact trustworthy. 'If a man [employed by us] has our complete confidence', he once said, 'we also trust him in his private

Figure 9.12   B.J. Edwards, Pye's Technical Director, 1958

affairs'. This attitude and his dislike of emotional confrontations of any kind made it unlikely he would summon B.J. for a timely chat about what had gone wrong with his life.

He took action at the last possible moment. Driving to London with B.J. in the autumn of 1959 he told him he was an alcoholic and said he had to stop drinking. Edwards paid no attention, and as Christmas approached the story went round Pye that he was having to pay the bills of Cambridge tradesmen with one pound notes; there was also gossip about a new office affair. In early December Sir Ben Barnett alerted C.O. from Australia to grave difficulties caused by the late delivery of equipment from TVT, the television transmission division for which B.J. was responsible. At a board meeting attended by B.J., C.O. warned that the size of their investment in Australia meant that 'the bad name the company has acquired could have a most serious effect on the rest of Pye'. He also complained that when one of B.J.'s assistants was told to provide information on TVT affairs in his absence, he had 'asked to be excused from giving any information'. C.O. said this was 'an intolerable situation which could not be allowed to continue'.

Almost as embarrassing was the discovery that B.J. had obtained a large order for equipment from ATV by quoting unrealistic prices and delivery dates. ATV had also asked him to investigate the possibility of pay TV, only to be told not to worry because he was working on a much better system. This much better system was never more than an idea in his head. Further enquiries revealed he had borrowed money from Pye representatives in Canada and the United States, telling them that he had cancer and was going to the Mayo Clinic, or that his wife had inherited a fortune and he would soon pay them back.

C.O. saw B.J. in his office on 14 December and told him he thought he was in trouble. According to C.O.'s notes of the meeting B.J. at first denied anything was wrong.

> I then told him that for years he had exaggerated and not told the truth and that he could not continue to run a business in this way. He told me that five years ago he had had a shock and that, since then, he was drinking. He referred to his private life but I refused to discuss the matter.

B.J. broke down and said he had to resign 'as he had done the company considerable damage'. C.O. replied that no one was asking for his resignation, and there the matter rested. Shortly afterwards C.O.'s Cambridge doctor Edward Bevan diagnosed B.J. as manic depressive and sent him to a psychiatrist who recommended putting him 'on some inventive kind of work near to Cambridge, where he could be kept under observation, and that his marital affairs should be cleared up'. A committee under Charles Harmer set to work to clear up the mess at TVT. 'The lady involved in the case' (C.O.'s expression) left Pye.

C.O. gave B.J. 6 months' leave of absence during which he was forbidden to have any contact with TVT. He also sent his Cambridge lawyer Walter Kester to Australia to examine the possible legal consequences of B.J.'s mismanagement there, and in early January wrote to Kester to warn that if Edwards turned up he was neither to be given money nor allowed to sign any document in Pye's name. And there was to be 'no drinking with the patient'. A board meeting called at short notice on 26 January 1960 heard Fred Keys report on the result of the investigation into B.J.'s expenses that C.O., after long hesitation, had allowed him to carry out. The surviving minute of this embarrassing meeting says only that 'a general discussion took place on the position of Mr B.J. Edwards who was still on leave of absence'. B.J. died on 16 February after taking an overdose of sleeping pills mixed with alcohol. When C.O. heard the news he telephoned his office at Pye and ordered the destruction of all copies of Keys' original minute of the emergency meeting that had probably ended in a call for Edwards' resignation.

The tragedy reflected badly on C.O. He had long worried that B.J. might spend too much money on schemes that would not work, yet never kept a check on what he was up to. Had it not been for his queasiness in personal matters he might have faced the problem earlier, and at a less terrible cost to all. Yet when the crisis did break he treated his old associate with compassion, and later paid for the education of his children. If he clung to an evidently damaged man too long it was partly because they were both originals, intolerant of boredom, and shared the same dangerous belief in the invincibility of their superior intelligence.

C.O. never tried to replace B.J. Edwards at Pye; perhaps the company was now too big and too fragmented to allow any one man, however gifted, to give it technical edge as he had once done. That C.O. seldom if ever talked about his old colleague was less surprising. As he had shown in his handling of the crisis of the VT 14, silence was his way of 'never giving in'.

A short economic boom in 1959–1960 made it easier for C.O. to ignore both the shortcomings of his technical staff and the gap left by B.J. Edwards' death. The government's removal of restrictions on hire purchase and TV rentals in the autumn of 1958 triggered a surge of demand the following year. An unprecedented 2.75 million television sets were sold in Britain, and the Pye group achieved new record profits of £2.4 million. C.O.'s attention was also distracted by two favourite hobby-horses – commercial radio, and the backwardness of the 405-line standard still used by British television although most of Europe had switched to the superior 625 lines. Both issues were given added urgency in 1960 when the government appointed the Pilkington Committee to make yet another report on the future of British broadcasting.

C.O.'s pioneering passion for commercial radio was well known, and according to his ATV colleague Ellis Birk he got 'incredibly frustrated' when people did not understand or would not listen to his monologues on the subject. John Stanley shared his father's passion, which Peter Threlfall interpreted as part straightforward desire to sell more radios, part 'intense dislike of [Britain's] ... over-regulated, mandarin-controlled' broadcasting system. When the Royal Show

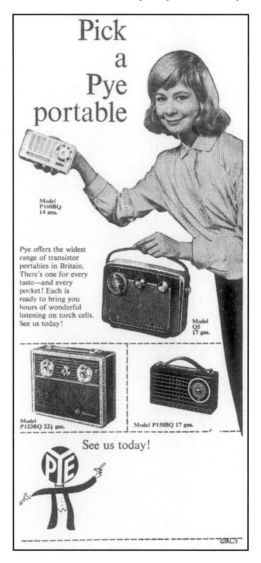

Figure 9.13   Advertisement, 1959

came to Cambridge in the summer of 1960 the Stanleys exhibited a
local radio station complete with studios and transmitter. Later that
year Pye published 'A Plan for Local Broadcasting in Britain' and a
map of towns with a population of 50 000 and over that C.O. thought
suitable for a commercially financed station. C.O.'s main point was
that only truly local broadcasting could satisfy local needs. 'Nobody in
Cambridge is interested in . . . traffic jams in Bedford. The Cambridge

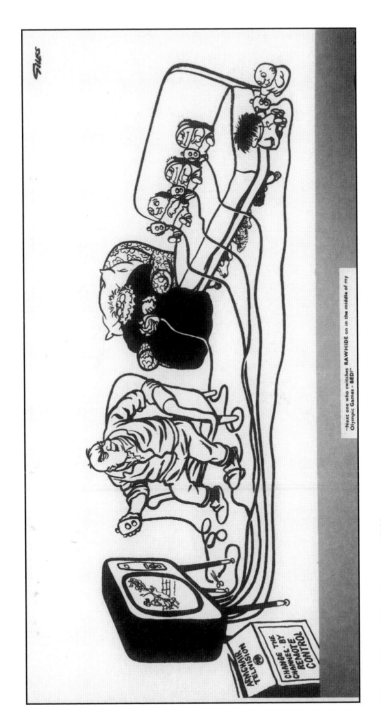

Figure 9.14   Giles Cartoon, 1960

housewife has not the slightest interest in what is selling in Bedford shops though they are only twenty five miles away.'

The Pilkington Committee's rejection of the Pye plan inspired the Stanleys to continue the campaign by other means. John introduced C.O. to Ross Radio Productions which owned the British rights to 'People Are Funny', a popular American radio show based on audience participation. Pye sponsored the show for 4 years and put it on in towns all over Britain as well as broadcasting it over Radio Luxembourg. This served the double purpose of letting the British see how commercial radio programmes were made and giving publicity to Pye, whose name, with the names of its local dealers, were displayed wherever the show was recorded.

Preparations for the launch of the British pirate radio station Radio Caroline in 1964 gave C.O. the chance to become involved in a more effective demonstration of commercial broadcasting. Radio Caroline evaded the Post Office's grip on the airwaves by broadcasting from a ship outside Britain's territorial waters. Determined that it should succeed, C.O. and John agreed to provide all its equipment, but because the operation was illegal they set up a fictitious radio manufacturing company complete with its own stationery, invoices and bank accounts. Everything Pye made for Caroline was stamped with the logo of the nonexistent manufacturer. In the year of Caroline's debut the Stanleys succeeded in opening on the Isle of Man Britain's first legal commercial radio station. Their partner was Richard Meyer, who had been involved in the start of ATV and then gone to Mozambique to set up, with Pye equipment, Radio Lourenço Marques, the world's first pirate broadcaster targeted at South Africa. John Stanley conducted prolonged negotiations with the Post Office to extract suitable frequencies for the new Manx Radio, whose advertising revenues put it in profit within 2 years.

C.O.'s campaign for the 625-line standard was equally ambitious. It offended his pride that Britain, which had once led the world in television, should now so obviously lag behind. And he calculated that the changeover to 625 would lead to a rise in TV sales comparable to the surge set off by the start of commercial television, thus bridging the gap between the present saturation of the market and what he was sure would be the golden era of colour television. Reappointed to the Television Advisory Committee (TAC) in 1959 for another 3 years he convinced its chairman that this was the most important issue before them, and TAC recommended to Pilkington that Britain should begin a gradual changeover to 625 lines in spite of the opposition of other manufacturers. With the 1959 boom dying away the latter feared that a change would leave them with large stocks of 405-line sets that could not be sold. They also knew that making dual-standard sets was expensive and technically difficult.

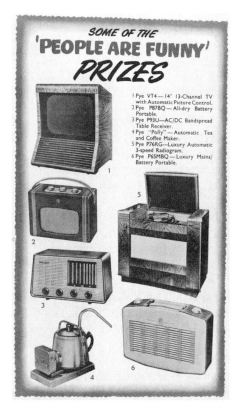

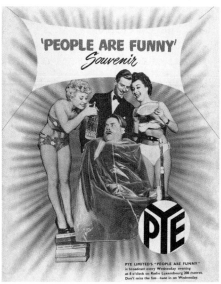

Figure 9.15   Souvenir leaflet, showing prizes offered, from the Radio Luxembourg programme, 1964

Figure 9.16   Installing Manx Radio, Isle of Man, 1964

In December 1960 the Radio Industry Council representing BREMA and the component manufacturers advised Pilkington to stick with 405 lines, and C.O. again walked out of the Association having only just rejoined it after an absence of 7 years. Rieck thought he welcomed the quarrel, and that his strategy was consciously destructive. By 1960 Pye was losing the sales battle against rivals such as Thorn and Sobell, so C.O. took up the campaign for 625 lines

> hoping (according to Rieck) that this would render obsolete every exist-ing TV set, cause everyone to buy a new dual-standard set, knock the big rental companies for six by destroying the value of their [405 line] sets, and create a paradise for Pye.

If this was C.O.'s plan, small wonder he decided to boycott the 1960 Radio Show and instead hired the Royal Festival Hall for a week-long solo exhibition. Pye's display included not just television sets and radios but products from all over the group, a surreal mix that included the genteel Hostess heated trolley and B.J. Edwards' wire-guided anti-tank missile that C.O. built as a private venture but could never sell. To make sure of the best possible press coverage he asked the designer

Alan Bednall to imagine what a three-dimensional television set of the future might look like. Bednall produced a globe of clear plastic 2 ft 6 in in diameter, furnished it with little figures on a surface made of plasticine and brought the picture to life with ingenious lighting and a tape-recorded soundtrack. The result was so convincing that C.O. got his headlines, and for several weeks was pestered by representatives of the Soviet trade delegation, who believed they had seen a real three-dimensional television and wanted to put in an order.

The success of the show at the Festival Hall and C.O.'s belief that he was about to outmanoeuvre his competitors over the new line standard may partly explain why he took his next step. At the end of October 1960 he told his board that he planned to bring Pye and Ekco together in a holding company called British Electronic Industries (BEI). This would only be a 'letter box', he explained, and the two companies would continue to operate 'independently in every way'. BEI's lack of substance was made plain by its board, which had C.O. and Cole as chairman and vice-chairman supported only by Eddie Stanley and the ageing L.G. Hawkins from Pye and one director from Ekco. There is no evidence that Pye's own board discussed the deal while it was in the making, but Fred Keys noted that all the directors present for C.O.'s announcement 'unanimously agreed that this was an excellent arrangement for the company'.

It is hard to see why. When C.O. telephoned E.K. Cole 7 weeks later to tell him that Pye's shareholders had accepted the offer of an exchange of Pye shares for new ones in BEI, Cole was in a crisis meeting with his Ekco directors. Stocks of unsold television sets had soared, the company's overdraft had reached £2 million and its subsidiaries in Australia were losing money. Ekco's results for 1960–1961 would show a trading loss of £432 000 on a turnover of £13 million.

Ekco was only a third the size of Pye, but 70 per cent of its turnover and profits came from the unreliable TV and radio market. The Ekco factory at Southend employed 5500 workers and its modern tooled production line made Pye's Lowestoft plant look antique. But Ekco had never regained its pre-war eminence as a maker of good quality receivers and its reputation sank further when one of its latest models, a supposedly portable plastic television set, proved almost impossible to sell. Matters were made worse when the receiver market collapsed in early 1960 and Southend went on turning out hard to sell sets until Eric Cole ordered a belated halt.

Did C.O. know he was buying what Peter Threlfall would later call 'a dud business'? True to form he did not allow his auditors Howard Howes to carry out a preliminary investigation of Ekco. Walter Meigh prepared a long list of questions he thought needed to be answered, but C.O. brushed them aside. He told Meigh he knew about Ekco's unsold sets, and argued they would prove a blessing when the receiver market

Figure 9.17   Futuristic, three-dimensional television receiver, produced by Pye at their exhibition at the Royal Festival Hall, 1960

boomed again and Pye needed additional capacity to meet the demand. He also ignored Meigh's objections that Pye was paying too much for Ekco, and that its Southend factory was making a loss. The merger, he told Meigh, would allow him to get a better price when buying cathode ray tubes and other components from Mullard, and anyhow it was not a merger but a disguised takeover, which meant he was getting Ekco cheap.

C.O. had known Cole for many years; their two companies had co-operated over airborne radar and other military equipment during the war. He trusted Cole and would not have thought it necessary or proper to check up on any figures he gave him. 'However misguided', commented Howard Howes' Michael Nathan, 'it was still the era of "my word is my bond".' At the same time C.O. never had a high opinion of the business ability of the quiet and rather secretive man whom in private he called 'Little Eric'. His opinion had further to fall. In January 1961 C.O. was obliged to tell the Pye board that the situation at Ekco was 'worse than anticipated', and he instructed Keys to estimate how much money would be needed to pay the first dividend to BEI shareholders. He drew up a nine point 'action plan' for Cole, and left for Australia to sort out the losses of Ekco's companies there.

Contacts between Ekco and Pye were uneasy from the start and there was no attempt to co-ordinate their activities. Nevertheless few who knew C.O. – apart, that is, from Cole himself – could have supposed that he would let Ekco go its own way for long; Cole had backed out of earlier merger talks with Jules Thorn when the latter made it plain he intended to take command of both companies. C.O. proved no different. He angered Cole by not sending him the figures for Pye's television and radio sales, though as his vice-chairman he had a right to them. Cole went to Cambridge to demand the figures shortly before BEI's first annual general meeting, only to be told by Fred Keys that he did not have them. C.O. then informed Cole he intended to run Ekco himself. After a brief resistance Cole agreed to stand aside, but refused C.O.'s offer of a golden handshake on the grounds that the company could not afford it in the present hard conditions. The dignity of Cole's departure was somewhat spoilt when a few months later he asked C.O. to sell his company back to him. C.O. said no, but it was obvious that Cole lacked the finance for any such deal.

C.O. behaved as though his vice-chairman's resignation had solved the Ekco problem. Writing to his Australian friend Arthur Warner in November 1961 he complained that it had been impossible to 'deal with the situation [at Ekco] because one was forced to work through Eric Cole'. Now Cole was gone he had 'a really fine business' that included the 'first-class' factory in Southend with a capacity to make 10 000 receivers a week. By C.O.'s own admission, though, the Southend factory was only producing a fraction of this number. Warner had been forced to take harsh measures by an even sharper downturn in the Australian market, and it is hard to believe he would have applauded the merger with Ekco. To an informed onlooker such as Michael Nathan it made no sense at all. Why had C.O. gone back on his public undertaking to extricate Pye from the treachery of the receiver market? A merger with a company whose profits largely depended on domestic sales of radio and television seemed impossible to justify.

Nathan would have been enlightened, but scarcely encouraged, had he heard C.O. explain to a joint meeting of Pye and Ekco directors in April 1961 the circumstances in which he and Cole had struck their deal. They reached their preliminary understanding, he said, in June–July 1960 when Pye was preparing for its 'very successful' solo show at the Festival Hall. There was a 'feeling of optimism' throughout the company and an 'assumption that Ekco felt the same'. When the merger documents were signed 'one felt stock [was] moving very slowly and then the market fell away completely and sales were only sixty per cent of forecast'. It was a poor piece of self-justification. June 1960 was the month Cole returned to Ekco from convalescence abroad after breaking a leg skiing and at once ordered the cutback in Southend's receiver production, a cutback some at Ekco thought already 6 months overdue. Had Howard Howes been allowed to take a closer look at Ekco it would have scarcely been possible to 'assume' that Cole shared C.O.'s bullish mood. Only in 1963, when overall profits for the group were recovering and BEI had been renamed Pye of Cambridge, was C.O. ready to offer a brief admission that something had gone wrong. His problem, he commented privately, was that shareholders would want to ask questions at that year's annual general meeting about Ekco's contribution to group profits 'while really we only want them interested in the integrated figures [for the whole group]'. Accordingly his formal statement said no more than that

> for some years prior to 1960 the Pye company had been working to maintain the right balance between the percentage of turnover and profit depending on consumer radio and television goods and the turnover and profit from other sides of the business. However, with the formation of BEI in 1960, the balance was tilted in the wrong direction.

It is striking that he never boasted, as he could have done, that after the merger BEI had 25 per cent of the British television market. That market, though, was stagnant, and within four years C.O.'s share of it would fall back to a modest 15 per cent.

This difficult turn in his affairs put further stress on C.O.'s flawed relationship with his son. Those who knew John Stanley had always doubted his father's wisdom in moving him in 1955 from the management of Telecomm to running merchandising for the whole of Pye, a task for which he had neither experience nor proven aptitude. C.O. further increased John's responsibilities when he made him Pye's deputy managing director in December 1961. John was now trapped in the consequences of the Ekco merger but, his new title apart, had no greater purchase on his father's actions.

Robert Browning had not been impressed by the boss's son when he first arrived at Pye in 1959. He noted John's small chin, the 'Peter Wimsey lock of blonde hair falling across his eyes' and his loud voice. He thought him 'arrogant and self-opinionated', and lacking in respect for his father in the way he walked into C.O.'s office, without knocking, to interrupt whatever conversation was going on. On longer acquaintance Browning changed his opinion, but even close friends admitted John was difficult to know. Shyness could make him abrupt, while behaving as C.O. expected of him contradicted his own gentle nature. C.O. had an easy knack of rallying people behind him; John did not. He had his father's forthrightness, but not the gift of charm that stopped it grating. It did not go down well with the directors of ATV when he reproduced his father's hectoring tone while standing in for C.O. at board meetings,

In some ways son and father were alike. John had inherited C.O.'s early Thatcherite genes: he was just as committed to free enterprise, just as determined an enemy of government regulations and monopoly. But that was not enough for C.O. Fred Keys, so loyal to C.O., told the story of the one and only visit John's father-in-law made to Pye. When his tour was over he turned to John and said, 'You will never be happy working in the same business as your father, go and find your own business to run. If you stay at Pye you will regret it.' If John had doubts about working for C.O. – and there is no evidence that he did – they were dispelled by the engrossing years at Telecomm. When C.O. decided to move him on from there in 1955 it was too late to break away.

Even Keys thought C.O. had 'bullied' John into an unsuitable job when he put him in charge of Pye's merchandising. John did not inherit the same instinct for the market that helped his father through earlier crises. John Gorst, who knew John well and liked him, soon saw he lacked C.O.'s skill at catching onto news headlines to boost Pye products. When it came to deciding on new products, John could throw reason to the winds and rely on instinct and a decision made off the top of his head, which Gorst knew was not C.O.'s way at all. And John's promotion brought no end to his father's interference in his work. In 1963 John decided it would be good publicity for Pye to sponsor a day of racing at Newmarket. The meeting was to start with the Invicta Stakes; there were also to be the Pye Stakes, the Ekco Nursery Handicap and, in honour of the elusive 625-line standard, the 625 Handicap. C.O. had gone to Ireland for his summer holiday but let it be known he thought the race meeting a bad idea, and refused to say if he would fly over for it as John hoped. When he did not show up at the opening ceremony it was assumed he had stayed at Lisselan, but later that afternoon he was spotted drinking in a race course bar with Norman Twemlow whom he had summoned to keep him company. He was just as unco-operative when his son tried to find suitable nonexecutive directors to strengthen the Pye board. John consulted about candidates with Renwick and with Anthony Burney, one of the City's best-known accountants. C.O. was not interested.

The ill-timed merger with Ekco brought about a collapse in group profits to £727 000 in 1961 and £118 000 the year after. C.O.'s first response was to cut back. Citing the example of Arthur Warner's cold-blooded reaction to the slump in Australia he demanded cost cuts (including a 2-year freeze on executive salaries) and the closure of all loss-making operations. He even raised the possibility of job losses, though throughout his life he had preached the employer's duty to protect his workforce come what may. But he did nothing to reduce the number of his own three receiver trademarks (Pye, Pam and Invicta) now augmented by 'Little Eric's' Ekco, Dynatron and Ferranti wireless and television. They continued to go their own way even though they were virtually competing with each other.

More positively, he pursued his belief that the future of British television lay with the 625-line standard, and in his evidence to the Pilkington Committee argued that only the stimulus of technical change could revive the television market, calling 'short-sighted' the fear of other manufacturers that they would be left with stocks of unwanted 405-line sets. He added that Pye, as Britain's largest exporter of television sets, was already making only 625-line sets for foreign customers. At the same time he readied Pye for a changeover in Britain by ordering production of a dual-standard set that operated on both systems. This again put him at loggerheads with his colleagues in the industry who wanted to ban 625-line sets from the 1961 Radio Show. C.O. stood his ground, and Pye and Ekco demonstrated a dual-standard receiver whose tuner had been designed in Germany. Other makers could only show 405-line sets they claimed could be adapted to the new standard. C.O. had apparently trumped the trade again.

His triumph seemed complete when the following year the Pilkington Committee recommended that the BBC provide a new programme on 625 lines (the future BBC 2, which began broadcasting in 1964). It also ruled that BBC 1 and ITV, while continuing to use 405 lines, would duplicate their programmes on 625. In fact the new standard posed a problem for Pye. Dual standard sets were virtually two television sets in one, technically more challenging to make and also £10 more expensive than an average 405-line set.

This put Pye at variance with the new trend in the television business – rentals. It was becoming more popular to rent a television than buy it, not least because of the chronic unreliability of sets and the expense and inconvenience of having them repaired, and the rental companies demanded the cheapest possible equipment. C.O. largely ignored this development. Richard King, then working on the commercial side of Pye, judged him to be still 'passionately against the rental syndrome . . . Rental to him meant the whole thing was going into the hands of the financiers who were at odds with the High Street trader whom he loved and respected.' C.O. resented the rental companies' exploitation of their size to buy sets from manufacturers at prices £10 to £20 cheaper than those paid by dealers. Other manufacturers were forming their own rental companies to get round this problem, but C.O. was already heavily borrowed and he would not go to the City for the money needed to finance an operation of this kind.

The growing popularity of rented sets was discussed at Pye board meetings from 1957 onwards. In 1958 C.O. calculated that half of Thorn's sets were bought by rental chains, allowing Jules Thorn to avoid the traditional ups and downs of the television market and install automated production lines. While Thorn invested in his own rental company, DER, C.O. stuck with the shops, whose business was diminishing. The man who had always prided himself on staying ahead of events was now immobilised by loyalty to what he still thought of as the 'family' of Pye dealers. His success at Pye dated back to the time in the 1930s when he took distribution of radios away from the wholesalers and gave it to the independent dealers. Since then he and Eddie had nursed and nurtured them in the belief that they held the key to Pye's success. It was the sort of personal business-making in which C.O. took delight and excelled.

In 1962 John Stanley reported to the Pye board that as many as 4 out of every 5 television sales were to rental firms, adding that though this proportion might diminish it would not drop below 50 per cent. There was no longer any doubt that Pye, burdened with Ekco's excess capacity as well as its own, had to assault the rental market, but there were two problems. Most of the decent rental outlets had been bagged by rivals. And no one in Pye had experience of running a big rental operation. As far back as 1958 C.O. had talked vaguely about the need to put one man in charge of rental business. 'With the benefit of hindsight', Fred Keys admitted, 'we should . . . have recruited a senior man from Thorn or Radio Rentals, but it had never been Pye's style to recruit senior people from outside.'

If ever there was a tar baby it was Pye's venture into the big-time rental business, and C.O. gave it to John. That was his first mistake. The second – his failure to keep check on the financial implications of John's activities – followed inevitably from C.O.'s style of management. In 1960 he had sent John Brinkley, Telecomm's new managing director, a note on what his management priorities should be. C.O. told Brinkley that success depended on four things – imagination, factory, finance and sales – but that he should only bother about the first and the last, in other words the two areas where C.O. himself excelled. The factory Brinkley 'should be able to forget completely'; finance could be left to a 'competent' person. It was a snapshot of how C.O. had run Pye, with great success, for 25 years. In the rental business, where success depended on low costs and tight financing, it was an invitation to disaster.

John's first sortie into the rental market followed C.O.'s instruction to help those Pye dealers who were threatened by falling sales but capable of renting sets profitably. He came up with Check Rentals, a co-operative rental system financed by Pye and first tried out in 1963 in Northern Ireland. It succeeded there, perhaps because the dealers who joined the scheme included both Protestants and Roman Catholics and in this instance the wayward chemistry of Ulster bonded them together. Transferred to mainland Britain it flopped. Michael Rose, the young solicitor so awed by his first sight of C.O. and now a partner in Pye's lawyers Bartlett and Gluckstein, helped launch the operation. Pye would invite local dealers to a big dinner with plenty to drink. Rose then explained how Check Rentals worked, and was followed by a Cockney salesman who bullied the well-oiled retailers into signing up. Richard King thought the idea flawed from the start, a 'high cost operation' with 'incredible financial implications' for Pye, and 'set up in the [erroneous] belief that there was enough margin [in rentals] for two profits'.

At its peak Check Rentals rented out sets at a rate of 1000 a week, a fraction of what Pye needed to remain in the television business. John's urgent search for more outlets led him to Donald Gibbard, a Bristol radio retailer who had first caught Pye's eye in the early 1950s when Eddie Stanley was building up dealerships in the south west. Thanks to generous credit from Cambridge, Gibbard had acquired by end of the decade a chain of 20 shops without C.O. noticing anything odd about him.

C.O. drafted in Norman Twemlow to work in tandem with his son. A director of Pye since 1951 and an effective salesman of the old school, Twemlow had no more experience of rental operations than John. His recommendation to C.O. was loyalty (people said of Twemlow that he was 'in C.O.'s pocket') but even old colleagues such as Keys and Charles Harmer thought it wrong to put him to work beside John in such a tricky operation. John also had help of a different kind from Cecil Rieck, the émigré from Nazi Germany whom C.O. liked to describe as 'a very clever man who had made a lot of money for himself'; cynics said Rieck built his fortune by buying shares on his own account two days before putting in a bigger order for C.O. Pye's professional advisers at Howard Howes and Bartlett and Gluckstein found Rieck sardonic, even sinister, and were never sure what he was up to. In a company with little time for titles Rieck, if pressed, described himself as executive director with responsibility for financial negotiations, and it was in this capacity that he began the discussions that led to the transformation of Donald Gibbard from Bristol radio dealer into big-time rentals operator.

The first hint of danger came when Rieck fell out with Gibbard and withdrew from all further dealings with him. John shared his father's admiration for Rieck, but he did not let the incident upset his plans. As Michael Rose explained, Pye was drowning in a 'tide of desperation' over the collapse of the television market and 'Gibbard seemed to hold out the key to success'. With much help from Pye, which supplied him television sets on tick and guaranteed his credit with banks and finance houses, Gibbard bought up a series of small rental chains, and eventually amassed some 150 shops. At the beginning of 1963 Pye's commitment to the Gibbard group was less than £1 million. Three years later it had grown to over £5 million.

C.O. was not involved in the details of the building up of Gibbard, and since Pye kept the relationship secret in order to avoid upsetting its old dealers he was never challenged on it in public. But he did follow the deterioration of the Gibbard group's financial position. In the summer of 1963 Gibbard issued to Pye £565 000 convertible loan stock which represented money owed to Cambridge and, if converted, would have given Pye 49 per cent of the rental group's equity. This deal to turn Gibbard into the lynchpin of Pye's rental operations was a cause for general self-congratulation at Cambridge, where Eddie Stanley invited him to stay at his house, and Gibbard soon became the guru to whom Pye turned for enlightenment on the mysteries of television rental. C.O.'s insistence on keeping quiet about his rental activities meant that the 1963 arrangement 'unfortunately', John later admitted, did not give Pye control of Gibbard, though he thought it 'did allow [Pye] to be fed with all [Gibbard's] information . . . as if we had been a director of the company'. He had not reckoned with Gibbard. When Michael Rose was sent to explain to him that as part of Pye's financing he needed to issue a debenture he would not even let the lawyer into his office.

It took 12 months for John and Twemlow to understand how bad Gibbard's accounting and auditing were, though the warning signs of poor administration, shortage of cash and questionable accountancy were evident by 1963. At that time Rieck was still involved with Gibbard and warned C.O. that Pye was being drawn into the 'old story of someone who had run a small business successfully without nor-mal accounting information, believing that the same policy will serve when the business is much larger'. But it was too late to draw back, for Gibbard was on the way to becoming the most important single customer for Pye television sets. By the time Pye prepared its report on the financial year to March 1964 it was already unlikely it would be able to recover all the money owed it by the rental group, though the accounts then signed by C.O. made no mention of this.

Figure 9.18   Loading 625-line transmitting equipment for BBC 2, Lancashire, 1964

1964 brought other problems. The BBC's new 625-line channel that opened in April aroused little general interest; BREMA called the public's reaction 'apathetic'. For C.O. it was the usual story of mandarin broadcasters giving the people what they thought was good for them rather than what the people wanted. When Labour won the October 1964 general election C.O.'s margin of manoeuvre was further reduced: lobbying Conservative governments on television had been hard enough, but he could expect no sympathy from the new prime minister Harold Wilson. He went into battle nonetheless, and in December, accompanied by Jules Thorn and BREMA's director S.E. Allchurch, called on the new minister of technology Anthony Wedgwood Benn. It was not a meeting of the minds. Benn noted in his audio-diary that his visitors were 'extremely disappointed' that the arrival of BBC 2 had not stimulated TV sales

> on the scale they needed to bring the industry up to full capac-
> ity...What they want is a second entertainment [i.e. commercial]
> channel to create a mass demand for them. I found them an extremely
> unimaginative lot. Frankly we cannot reach a decision of the kind they
> want just to give them work. What is required is an assessment of the
> unused capacity and then a decision how best it can be used in the
> national interest...If these were the leaders of British industry in this
> field, they were a poor lot.

Figure 9.19   Pye 625-line school for service engineers, 1964

C.O. also suffered a disappointment that he would not readily admit. While the Conservatives were still in office in 1964 Robert Renwick received one of Britain's last hereditary peerages for his work in the war, his campaigning for private enterprise, and years of fund-raising for the Tory party. Out of gratitude for C.O.'s contribution to these achievements Renwick enquired about an honour for his old friend. The reply, he told C.O., was positive, but qualified. A knighthood, perhaps even a peerage, could be arranged, but on condition that C.O. made a large contribution to a Conservative party industrial pressure group. He thanked Renwick, then said 'Haven't I done enough already?' His idea of honour did not allow him to buy what he believed he deserved on merit, and there he let the matter rest.

By 1965 even the largest and most efficient rental firms such as Radio Rentals and Rediffusion were facing difficulties. The Labour government had tightened credit restrictions in a market where expansion was already coming to an end and rental income falling. A late-comer to the rentals feast, Gibbard would have found the competition hard enough had his business been well managed but, as John Stanley gradually

Figure 9.20   C.O. Stanley with Lord Hill, Chairman of ITA, at the National Radio Show, Earls Court, London, 1964

discovered, this was not the case. After Gibbard had moved his financial headquarters twice in less than two years 'it was obvious (John's words) that this was bad management and must cause loss and confusion which in turn must affect the [company's] figures'.

Figure 9.21 *Princess Margaret with John Stanley at the Pye factory, Cambridge,*
1963

Gibbard's questionable figures featured prominently in Pye's discussion of its own accounts for 1964–1965. Norman Twemlow assured
Howard Howes at the beginning of August 1965 that the Gibbard
group's indebtedness to Pye was covered by the value of its rental and
hire purchase agreements. Twemlow sent a copy of this letter to John

Figure 9.22   Princess Margaret talks to Donald Jackson during her visit to the Pye factory, 1963

and Keys, and neither made any objection. Other Pye directors were uneasy about the way John and Twemlow kept the Gibbard dossier to themselves. In August Rupert Jones, who was running the Ekco plant at Southend, told Fred Keys that Gibbard's problems 'should be brought before every regular board meeting until we are satisfied that the business is proceeding satisfactorily'. He sent a copy of his letter to Harmer, whom he knew also felt that the board should 'be kept more fully informed on this issue than has hitherto been the case'. Eight days later Walter Meigh of Howard Howes told a Pye board meeting of his 'great concern' over the Gibbard debt, which now stood at £5 million, plus another £1 million in guarantees. Meigh advised the board to make a provision for this in its accounts but John, who was chairing the meeting in C.O.'s absence in Ireland, insisted he and Twemlow could tidy up the rental group's affairs without any loss to Pye. According to Keys' minute not one of the directors spoke up in support of Meigh, and all voted to make only a £1 million contingency reserve with no effect on Pye's profits, a decision largely based on Twemlow's underestimation of Gibbard's liabilities and overestimation of the value of its rental agreements, When the meeting was over, Meigh asked Harmer why he had kept silent. Harmer said, 'I've given up banging my head against a brick wall.'

Figure 9.23   *Prince Philip, Duke of Edinburgh, with C.O. Stanley (centre), at the Pye factory, Cambridge, 1964*

The brick wall was C.O. He stood behind John and Twemlow, and had he wanted proper provision for the Gibbard debt he only had to send word to Keys and it would have been made. The board's decision that day allowed him to claim Pye had achieved profits of £5 million, almost half from sales of radios and televisions at home and abroad, £1 million from Telecomm and £750 000 from Pye's instrument-making subsidiaries. He could not have claimed these results had he made the £2 million–£2.5 million provision against the Gibbard debt that prudence called for.

Within days of the board meeting Gibbard went back on his undertaking, extracted with difficulty by John, to give Pye 51 per cent of his company's equity as collateral for the debt to Pye. C.O. drafted a tough riposte pointing out that Pye's investigations revealed a situation in the rentals group that 'justified their worst fears', and demanded Gibbard honour the agreement. Gibbard buckled and John and Twemlow joined his board, with Twemlow as its chairman. The following month they reported that Gibbard's shops did not have enough rental agreements to sustain the business. Pye at last understood that the operation had been unprofitable throughout the past 4 years, and that a quarter of Gibbard's staff were underemployed.

Figure 9.24   Leaflet for dual-standard receiver, 1964

John knew his father's inclination to ignore problems in the hope they would go away, but C.O.'s behaviour in the summer of 1965 suggests he was now a worried man. One sign came when he asked Cecil Rieck to prepare an analysis of the decline of Pye radio and television sales over the previous ten years. Rieck's chronic scepticism and privileged position as C.O.'s truth-teller made it unlikely his report would make cheerful reading; in the event it took criticism to the verge of contempt.

He examined the poor design of recent sets and their high production cost, Pye's lateness in entering the rental market, and the 'political' campaigns such as C.O.'s propaganda for the switch to 625 lines. The latter, he argued, had alienated Pye dealers who were trying to sell their stock of old 405 line sets and encouraged the public to hedge against technical change by switching in ever greater numbers to rented sets. 'Pye [Rieck meant C.O.] believed all these unfortunate side effects were worthwhile for the millennium of [625-line broadcasting] to come,' but BBC 2 flopped as a popular channel, 'the boom never came [and] Pye has suffered all the odium for nothing'.

Rieck's recommendations were as uncomfortable as his criticism. Future policy should be 'properly thought out [and] followed with reasonable consistency'. There had to be no more 'gimmicks and stunts' because 'political coups can boomerang'. More attention had to be paid to efficiency of production. When Pye gave credit there should be proper controls and 'one [must not], everlastingly, find oneself having to grant absurd amount of credit because of the pressure of unsold stocks'. Television production capacity was too big, and should be reduced to a 'realistic size' for a future in which sales would be mostly for replacement, and no higher than 1.6 to 1.8 million a year. Efficiency demanded a reduction in the number of receiver trademarks.

Finally Rieck turned to Gibbard. With £6 million of Pye's money at risk in the rental chain there had to be a full assessment of Gibbard's affairs, something C.O., true to habit, had refused to sanction. The job should be given to an established firm of accountants with experience of the rental business such as Peat, Marwick, Mitchell and Co 'who can put their fingers on unsound basic methods of rental accounting, recognise immediately the distortions produced and re-cast the figures . . . in terms of prudent practice'. The problem of Gibbard, he stressed, was now 'too big . . . to hide'.

C.O. sent his answer at the end of September.

> I disagree with [your] report completely except on one issue: that the production price of television sets in the Pye company is not low enough to compete with some makes at this moment. [Your] history of the company's activities is of course quite incorrect.

He implied that Rieck's early quarrel with Gibbard was the original cause of the rental crisis and, as usual when an argument turned against him, countered most of his adviser's points with mischievous barracking. The cockiness was a cover-up. Two months earlier, apparently unknown even to Rieck, he had embarked on secret negotiations to save his company from disaster.

Sources

COS files: 2/14, 2/15 (Stanley's notes on showdown with B.J. Edwards), 2/17 (Pye annual reports), 2/1 (Rieck's paper on Pye), 2/2 (B.J. Edwards), 2/19, 2/21, 3/2 (Pye recommendations to Pilkington), 7/1/6 (1958 speech to executives on failure of the VT 14).
COS/TWEM2/1 (1958 survey of radio industry), COS/TWEM2/2 (Ekco).
Simoco boxes: Box 1 (Cooper Bros reports including Gibbard), Box 4 (ATV and B.J. Edwards).
Pye main board minutes, Ekco board minutes, 1960
Public Records Office: HO244/32 and 255/189 (Post Office); HO244/258 (Pilkington).
Interviews: Brian Beer, James Bennett, Derek Cole, Don Delanoy, Jo Fletcher, Dennis Fuller, Sir John Gorst, Donald Jackson, Fred Keys, Jim Langford, Marjorie McCarthy, Walter Meigh, Patsy Morck, Michael Nathan, Michael Rose, Dadie Rylands, Nicholas Stanley, Peter Threlfall, Daphne Whitmore.
Michael Bell interviews: David Fernie, Charles Harmer, Richard King, Sir Charles Powell, John Stanley.
A–Z files: Sir John Gorst (Robert Browning's recollections).
Books: Bell, *ibid.*; Benn, *Out of the Wilderness*; Cairncross, *Managing the British Economy in the 1960s*; Geddes and Bussey, *ibid.*; Owen, *From Empire to Europe*; Pollard, *The Wasting of the British Economy.*

Chapter 10

The palace revolution

The crisis that now unfolded was seen by many at the time as a humiliation for the Stanleys – the father who would be kicked out by the company he had created, the son subjected to months of public attack and innuendo. It seemed C.O.'s mistakes had caught up with him at last and his flaws stood revealed. The truth was, as the truth usually is, a good deal more shaded.

By the middle of the 1960s Britain's television manufacturers were, without knowing it, an endangered species. The first black and white sets to arrive from Japan at the end of the decade had a quality and reliability British manufacturers could not match, Japan's colour sets would reinforce that superiority in the years ahead. Three decades after Cecil Rieck sent C.O. his damning report on Pye, five out of every six television sets produced in Britain were made by Japanese or other Far Eastern companies. C.O. had already seen the Japanese challenge at close hand in Ireland. When the Irish government gave Sony permission to open a radio factory at Shannon in 1959, he warned the Taoiseach Sean Lemass that 'this Japanese disease' was destroying the electronics industry 'in one country after another'. C.O. knew he could not match the low prices and near perfection of the Japanese; he also knew that Pye, even without competition from Japan, lacked the money to develop the technology that was driving his industry forward.

In the midsummer of 1965 his sense of the dangers ahead led him to call on J.P. Engels, chairman of Philips Industries, the British arm of Philips in Eindhoven. Engels was not surprised to see him. Some years earlier C.O. had asked Philips for an assurance of help should Pye find it hard to finance the research and development it needed to stay competitive. Philips was the obvious place for him to turn: his association with the Dutch company went back 40 years and he knew, and was known by, the people who ran it. Philips was the biggest electronics group in Europe (by 1995 it would be the third biggest producer of consumer electronics in the world after Sony and Matsushita). If anyone could help Pye survive in some recognisable form it was these Dutchmen.

Engels may have known that C.O. had already talked with RCA of America about joint ownership of a factory to make colour television tubes. RCA lost interest, but C.O. remained convinced that Pye's future depended on its ability to exploit what he believed would be the boom in colour television. He had never had the money to expand Pye's own tube-making subsidiary Cathodeon, hence his purchase of black and white cathode ray tubes from Philips' subsidiary Mullard. Engels was better placed than anyone to know that Pye could not afford the even greater investment needed to make colour tubes.

It was a similar story with the transistors (semiconductors) that were replacing the thermionic valve to revolutionise the technology of radio. Pye was one of the first 25 international companies to take out a licence on this 1947 invention of America's Bell Laboratories. When C.O. sent Dennis Fuller to represent Pye at the seminar where Bell explained the transistor to the first licensees, he gave him a pep talk before he left.

> This, Fuller, is probably the most important development in our business since the invention of radio. The transistor will change . . . the lives of everyone in this country and we at Pye are going to be involved from the very beginning.

Fuller thought this a 'flawless intuitive judgement', and so keen was C.O. to get hold of the transistor that Pye was the first company to hand over the $25 000 licence fee to Bell. He did not ask anyone to look at the economic prospects of his investment, which given his company's perpetual cash shortage were not good. In 1956 Pye put out its first transistor radio under the Pam trademark, a sure sign C.O. was not yet confident of its quality. It was Britain's first portable transistor radio, but because Pye's subsidiary Newmarket Transistors was not fully tooled and its production costs high, the Pam 710 was twice as expensive as a conventional valve set. Other licencees started making transistors on a scale that Pye could only have matched by going into a joint venture but C.O., disliking joint ventures on principle, would not do that. He knew, though, that transistors held the key to cheaper television sets and continued to tell his directors that 'Pye must have a substantial interest in transistors to survive'. It was inevitable that C.O. should bring up both colour tubes and the new silicon (as opposed to germanium) transistors in his first discussion with Engels. He raised other matters too – the unprofitability of Pye's radio and television business, its difficulties with rental chains – but said he was confident of overcoming all these with Philips' technological support.

The danger was that Pye's profits would slip before Philips' help took effect, dragging down its share price and leaving the company vulnerable to a takeover bid from one of Philips' competitors. There had been

Figure 10.1   Front cover of a leaflet, 1956

rumours the year before that Rank Xerox would make a grab for Pye, and another scare when C.O. suspected Jules Thorn of quietly buying up Pye shares. Engels saw this as reason to move quickly, and he went to Cambridge to discuss matters further with the entire Pye board (the only director absent was C.O.'s recent appointee, the Australian Arthur Warner). The agreement they reached went far beyond the original proposal of technical co-operation, and it is likely it was always C.O.'s

Figure 10.2   Inside view of Pam's first transistor radio, 1956

intention that it should. He was now 66 and could only believe that, if he ever retired, John would need even more help than he in the difficult years before the arrival of the colour boom. What better solution than to seek protection under Philips' ample shade?

At Cambridge the two sides agreed that Mullard would supply Pye with all the cathode ray tubes and semiconductors it needed; that they would discuss the 'rationalisation' of their television production; that

Figure 10.3   On the beach: watching Pye's first transistor portable television receiver, model TT1 (operated on battery or AC mains supply), 1962

Philips would enter the telephone business by joining Pye in the Telephone Manufacturing Company; and that to 'safeguard the position in the meantime' Philips would start buying Pye shares, while Pye's directors and their families undertook, in the event of a takeover bid by someone else, to sell their holding in the company to Philips.

C.O.'s approach to Engels suggested a man who could still make a bold move to meet a correctly anticipated future, but more immediate events were slipping beyond his control. Some of his directors were exasperated by the Stanleys' secretive handling of the Gibbard tangle, and they had other grievances too. In the early 1950s C.O. had rewarded seven of his most valued staff by putting them into an investment company called the Septangle Trust which he later endowed with some of his own ATV shares. The lucky seven were Eddie Stanley, Charles Harmer, B.J. Edwards, James Dalgliesh, Peter Threlfall, Norman Twemlow and Fred Keys, and when Septangle (set up and managed by Cecil Rieck) prospered there was bitterness among those who were not part of it. Rupert Jones, a director of Pye from 1958, complained so often to John Stanley about the unfairness of Septangle that John, whom Keys thought 'much kinder and more sensitive' than his father, invited Jones to invest in a new private company of his own called Telephone

Answering Services. When it failed to produce quick profits Jones complained again and John bought back his shares at a generous price. Jones remained disgruntled, and was further upset when John criticised the poor quality of the television sets he was producing at Southend.

By appointing his son deputy managing director C.O. made him a target for the complaints that the discontented would never have dared direct at him. No one was more likely to be discontented than Brinkley, John's successor as managing director of Telecomm. It was hard to find anyone at Pye who liked Brinkley. He had the social assurance of a Cambridge tennis blue and was a first-class telephone engineer, but he was also (or so Dennis Fuller thought) 'full of bluster . . . a "Great I Am" . . . devious and crudely ambitious'. Colleagues told the story of Brinkley showing a photo of himself standing by his car outside No. 10 Downing Street where Telecomm was installing equipment. 'Look at that', he said. 'That's the company we keep'. Brinkley also felt bitter about his exclusion from the Septangle Trust and it was perhaps inevitable that he should see a chance for himself in John's difficulties with Gibbard. Peter Threlfall commented many years later that the fact of the Septangle Trust's existence 'guaranteed that those excluded would have their loyalty to the Stanleys sorely tested in times of trouble'.

C.O. soon learned about the murmurs in his boardroom. In November Charles Harmer pleaded with him to pay more attention to the group's instrument-making companies which for 3 years had been starved of capital while C.O. spent money on new acquisitions and propping up the rental business. Harmer warned him that staff were already leaving for other companies that had proper long-term programmes while Pye was trying to produce sophisticated instruments on largely obsolete plant, some of it dating from before World War I. He got no response from C.O.

The same month Rupert Jones wrote to all members of the board to argue, as Rieck had done, that, 'unpleasant though it may well be', outside accountants should be appointed to look into Gibbard. Brinkley and Harmer agreed with him, and the three wrote a letter supporting a suggestion made by Barclays that the bank meet with Pye's financial advisers Morgan Grenfell and its broker Greenwells to examine how the firm's liquidity might be improved. At a board meeting later in November Brinkley came to Jones' support when John Stanley attacked him for bad production standards at Southend, even though earlier he had told John that he agreed with him. In December Jones, Brinkley and Harmer openly criticised C.O. over the handling of Gibbard and again called for a meeting along the lines suggested by Barclays. C.O. refused, saying it would turn Pye's problem into 'a public matter', and instead made Rupert Jones financial controller, an appointment he can hardly have taken seriously given his contempt for financial watchdogs of any kind.

He was playing for time. His aim, he said later, was to 'avoid a show-down' in the Pye boardroom until he had sealed the agreement with Philips. Discontented directors counted for nothing when he was on the verge of making a deal that would save his company. And how could he see men such as Jones or Brinkley as a threat? They were his creations. Neither of them had built up a business of his own, the only test of worth C.O. believed in. And it was unthinkable that he might be 'betrayed' by Harmer, who had been in thrall to him since the distant days of Arks and *Radio for the Million*.

C.O. may not have taken his new financial controller's appointment seriously but Jones did, and quickly produced a damning survey of Pye's finances – and, implicitly, of C.O. There was 'an acute shortage of cash', with too many group companies losing money, while investment was often unproductive even in those that were profitable. Too much money was spent supporting sales of radios and televisions. The root problem was weak control over the group's resources, and Jones had begun working on a proper financial regime. This would take time because of Pye's 'great shortage of trained personnel for this task'.

What C.O. did not foresee was the gravity of the crisis into which Gibbard was dragging Pye. Shortly before Christmas 1965 John told the board that Donald Gibbard had agreed to take 6 months' leave of absence, and promised to have no contact with the business during this time. Pye had at last won control of the company it had kept afloat for the past three years, but if this was victory, it came too late. When Howard Howes examined Gibbard's books what they found was shocking even by the standards of this wayward rental chain. Gibbard's accounts department had not kept proper books for years and was still destroying records that were only 3 months old – a state of affairs, Walter Meigh warned C.O., that put Norman Twemlow and the other Pye-appointed directors who had taken charge of Gibbard at theoretical risk of criminal prosecution. Since no records at all existed for the period from October 1964 to May 1965, and very few up to September of that year, it was impossible to say how much money Gibbard had lost in its last financial year. As for the current year, Howard Howes expected a trading loss of around £500 000. Did C.O. know, Meigh asked, that in spite of leaving this mess behind him Donald Gibbard and his wife were still being paid a joint salary of £13 500 a year?

The new Gibbard revelations were of particular interest to Philips. That autumn C.O. had paid another visit to Jan Engels to warn him of possible losses in the rentals business that could 'seriously affect' Pye's own position, and to ask for help should this happen. The upshot was that by early January 1966 Engels and his colleagues had decided C.O.'s position was so much worse than in the summer that they needed a 'closer association' with Pye than they originally planned.

Figure 10.4   Pye's 'transistor train' at the National Radio Show, Earls Court, London, 1964

Figure 10.5   Pye's converted parcel train making a tour of the country to advertise the company's latest television receivers, radios and domestic products, 1965

C.O. had lost control of events, though a stubbornness hardened by age did not allow him admit it. On 10 January 1966 the Pye board accepted Philips' (now old) plan for equal co-operation in rental shops and the manufacture of telephones and television sets. The same meeting also saw the appointment of Jones and Brinkley as deputy managing directors and an agreement that Frank Duncan, managing director of Ether Controls, bought by C.O. in 1964, should join the Pye board. Unknown to C.O., Pye executives had for some time been saying to Duncan, 'this is a terrible mess. Can't we do something about it?' Jones, Brinkley and Harmer seem to have seen in Duncan a figure of sufficient distinction within the industry to give focus to their complaints, and to force the issue in a way they would not have dared on their own. C.O. saw Duncan quite differently. He was a little older than C.O., and had been a director of Murphy. He had also worked with C.O. in the early days of BRW (the radio relay firm in which Pye had a large interest) and served as chairman of the Radio Industry Council. C.O. called him an 'old friend', but otherwise had little in common with this orderly man who was a linguist and something of an aesthete. Above all he did not take Duncan seriously as a rival; he was just 'a pleasant character . . . who had never built a business'. If Jones and others wanted Duncan on the board, why not let them have him?

C.O. underestimated, perhaps knew nothing of, the extent of Duncan's alarm over the collapse of his own family fortune. C.O. acquired Ether Controls by an exchange of Pye shares when the latter were worth more than 20 shillings apiece and he was promising they would go many times higher over the next few years. By the beginning of 1966 Pye shares had fallen by more than half and Duncan and his wife, who owned Ether, had seen their wealth halved. If anyone had reason to want drastic action taken at Pye it was the Duncans.

The need for action became more obvious at the beginning of February when the Pye board recognised that a collapse in the profit margins of Pye's TV division was going to pull down the group's overall profits. There was also bad news from the Australian and New Zealand subsidiaries on whose good performance C.O. had heavily depended in the past two difficult years. His concern for Pye's indebtedness (a 1965 bank overdraft of £9 million, short-term loans of £7.6 million) can be judged by his determination to retrieve A$400 000 he had lent his friend Arthur Warner, even if only by transferring it to the Australian associate of a bank in Britain so 'we could get credit for this balance against our British overdraft'.

C.O. thought the situation in Australia grave enough to need his presence, and flew out there on 6 February. The day before he left he had lunch with Duncan and briefed him on his latest talks with Philips. The next day Duncan drove with C.O. and Velma to London airport; and

C.O. said goodbye to him believing Duncan agreed that no important changes should be made at Pye while he was away.

The stage was set for a scene known to any student of politics – the coup against the overconfident autocrat who at a time of crisis unwisely goes abroad. When the Pye board assembled for its first meeting in C.O.'s absence it made Duncan vice-chairman. Two days later it appointed David Hobson of Cooper Bros, Ether Control's auditors, as Pye's financial adviser. The title gave little idea of Hobson's power. From that moment the group's figures were under his control and not even Fred Keys had access to them. At the beginning of March Duncan called an enlarged board meeting at which Pye's accountants, brokers and Cooper Bros were all present. The meeting discussed Gibbard, but John Stanley thought its real purpose was to stage an 'inquisition' in front of as big an audience as possible. According to John's account, Hobson, sitting at Duncan's right hand, 'virtually called crooks and liars' the directors C.O. had appointed to the rental firm after Douglas Gibbard's departure. When John tried to defend them Duncan turned to him and said that Gibbard 'was not the only case of mismanagement' and that they would have to 'investigate all the shops in the group and [Pye's] other sales activities'. Unknown to John Stanley, Duncan had asked Cooper Bros to investigate Pye's affairs and the accountants, known at the time for their aggressive style, were probably already at work.

When C.O. got back to Britain on 8 March, the day the Pye board held its inquisition, the 'palace revolution' had gone too far to be stopped. C.O. was the first to use this term, which expressed his shock that an act of such gross *lèse-majesté* had been committed. Told by John what had happened in his absence he tried to see Duncan, but the latter fobbed him off with excuses. Even after this treatment C.O. failed to understand the gravity of the situation, just as some weeks earlier he had rejected Robert Renwick's suggestion that he should retire, and John announce a 'diplomatic illness' and take a holiday in Switzerland. Renwick had spotted the danger posed by Gibbard and the dissent at the top of Pye and, according to Keys, gently told his old friend this would be an honourable way out of the crisis.

C.O. would have none of it. Charles Harmer thought this was because he still believed he could hand Pye on to John, and certainly even after his return from Australia he behaved as though he had a trump up his sleeve. And in a sense he had. Philips was completing its plan for the 'closer association' Engels said was necessary because of Pye's ever more critical state. The chief Dutch concern remained that Pye might be taken over by a competitor after Philips had come to its aid with know-how and technology. Engels therefore proposed what might be called a tactful, pre-emptive takeover of his own. A new company would be created made up of Philips' rental operations and the Pye group in which Philips held 55 per cent of the equity. Pye shareholders were to be offered cash and ownership of 45 per cent of the new company. Pye itself would continue to 'operate independently', though Philips expected rationalisation of the two groups' operations.

Engels outlined the new plan to C.O. shortly after his return from Australia. C.O. said he was 'strongly in favour of it', but told Engels he could not put it to the Pye board while its members were divided. That, at least, was prudent of him, for there was little in the new plan to please those directors whom C.O. now called 'the conspirators', and perhaps nothing at all if C.O. continued to have the ear of Philips. On the other hand, if the conspirators got rid of the Stanleys they could hope the elderly and unambitious Duncan would soon retire, leaving Pye in their hands. The events that followed certainly suggest that was what they had in mind.

C.O. had to wait seven days before Duncan agreed to see him, and then giving him only 10 minutes of his time before the start of the weekly board meeting. The inelegance of Duncan's conduct may be explained by the embarrassment that is customary when someone is about to be stripped of power, though fear of confronting such a renowned fighter must also have played a part. But Duncan had numbers on his side. Among the seven directors present that day the Stanleys and the ever loyal Norman Twemlow were in a minority of three. C.O. tried, and failed, to have proxies represent the two absent directors, Arthur Warner and the former member of Ekco's board Derek Pritchard. The latter asked Renwick to represent him, but the canny stockbroker demurred. Warner's choice, Cecil Rieck, was too obviously partisan and at once rejected by Duncan's side. Only now did C.O. realise he should have brought more of his own people onto the board before leaving for Australia.

He went into the meeting having been told by Duncan that its purpose was to let his son choose between resignation and dismissal. After routine matters were dealt with Duncan got up and said John had to go. He blamed him for the rentals fiasco and recommended that Cooper Bros should investigate Gibbard. He raised the possibility that John himself might be 'suspended pending investigation', but also offered him a choice of resignation or taking long leave. The meeting broke for three days to give John time to 'consider his position'.

With the gravity of the situation now plain even to C.O. something like paranoia gripped the Stanleys. There were meetings at Sainsfoins and at Eddie's house. When Velma found Peter Threlfall's wife Betty staying with C.O.'s brother she wanted to drive her out suspecting, wrongly, that the Threlfalls had joined the plotters. Fred Keys' minute of the debate that took place when the board re-assembled was scarcely coherent, suggesting his agony at having to record such shocking events. There was an ill-tempered discussion about whether someone might want to buy Gibbard from them, followed by talk about the possibility of a takeover bid for Pye. C.O. claimed to know an American company that was keen to buy the group and added, perhaps to scare his audience, that 'having built up the business he valued it very highly but rather than have a divided board he would prefer a bid'. He made no mention of the latest deal on offer from Philips that was still known only to him. He defended John's work at Gibbard – 'no particular person was responsible for the present situation' – and said that if the leave of absence proposed for John was linked to suggestions of 'irregularity' at Gibbard he would advise him not to agree.

When the vote was taken only Twemlow and C.O. voted against John going on leave. The most painful of the votes in favour was that of Harmer, who had spent almost all of his working life with the Stanleys. John suspected an ambitious wife had given Harmer the idea that he could replace C.O. as Pye's chairman. Harmer himself explained that he had invested all his savings in Pye shares, and did not want them to become entirely worthless. A weak man torn in two directions, he knew he owed everything to C.O., but also saw the damage his old patron was doing to the company. His vote was the most telling sign of how far C.O. Stanley had gone astray.

It has to be assumed that throughout these days of crisis C.O. and his son met many times, though neither of them said anything about it. John seldom mentioned such painful matters. He spoke to his wife just once about his feelings for his mother and her difficult life, and the effort was so obviously agonising for him that Liz never brought up the subject again. C.O. would say that these months were a 'crucifixion' for him, but he was thinking of his own feelings, not John's. It is even possible he believed John was at least partly to blame for their misfortune.

Within days of John Stanley's humiliation C.O. was fighting for his own survival. On 28 March he took Cecil Rieck to Cardiff to talk to the new directors of Gibbard. That night Robert Renwick telephoned to warn that in 2 days' time the Pye board would meet to force him to resign or, failing that, dismiss him. His first reaction to Renwick's call was that the hostile majority on the board did not matter, and that he could still fight and survive, but when he went to see his lawyer on returning to London he learnt that Duncan had prepared two statements to be published simultaneously, one on his dismissal, the other detailing heavy losses at Gibbard. C.O. said the board's estimate of those losses at upward of £6 million was 'fantastic', and vowed to contest it. His determination crumbled when it was pointed out to him that an attack on figures prepared by his own board could only further weaken Pye's already feeble share price.

By chance or cunning his opponents had found his Achilles' Heel – Pye, the company that was his life, and with whose salvation via Philips he was now obsessed to the exclusion of all else. When he got to Cambridge on 30 March he was met by Duncan and his lawyer John Mayo of Linklaters, and the three of them talked separately while the board began its meeting. Mayo accused C.O. of accepting Twemlow's figures on Gibbard when he knew they were incorrect, and asked for his resignation. C.O. extracted an undertaking that if he resigned within 2 months he would not be dismissed, and then announced his decision to the board. He could not face speaking to Pye executives, who had been told to expect an important announcement, and as was his habit on embarrassing occasions sent Eddie in his stead. The mood darkened further when C.O. heard that Arthur Warner was dead. Prompted by C.O., Warner had cabled the Pye board asking it to delay all decisions until he arrived from Australia. His request was turned down, and he suffered a fatal heart attack shortly after.

C.O. later claimed he needed the 2 months' delay before the announcement of his resignation to give him time to conclude the agreement with Philips. Duncan's press release on Gibbard made that unlikely, for it prompted ever gloomier reports about Pye's health. It was no longer expected to pay a dividend, which the *Guardian* called a 'serious setback for a company already facing a cash problem'. The *Daily Mail* reminded readers that only the previous month Pye had had to buy off creditors with its ATV shares worth £2.5 million. It also reported that Pye shares had fallen to 9s 9d even though the company still had 10 per cent of the British television market.

Some years later a Philips executive recalled 'the total lack of financial control' at Pye at this time, and suggested that only extreme decentralisation had allowed the group to go on trading when it was actually bankrupt, for if the debts of all Pye companies had been consolidated 'no bank would have lent to them'. The *Economist* offered an epitaph that few who knew C.O. would have challenged: 'He was too good a salesman, convinced that there was no market so bad that he couldn't sell his way into it.'

Small wonder that on the day of such gloomy press comments C.O., ever mindful of what Philips would think, telephoned Greenwells and Morgan Grenfell to tell them not to let Pye shares stay below 10s. It seems he still did not understand the precariousness of his position, for he made the call in the presence of Duncan, who was furious and told C.O. he would now have to be dismissed 'for interfering in the affairs of the company by continuing to negotiate with Philips' (this was C.O.'s, and the only, version of the incident). Two months later Duncan rejected the Philips' proposal, calling it 'a back-door take-over of an important English group by a Dutch group, under an English disguise'.

On 6 April 1966 lawyers worked out the terms of C.O.'s departure at a meeting from which he, to his fury, was barred. His executive duties in Pye ceased at once, while he was to resign from all his directorships in the group at the end of May and assume 'the honorary office of president'. He was also given a 2-year consultancy at an annual salary of £10 500 and during that time the use of his Rolls Royce and chauffeur. Pye agreed it would not use his resignation to support allegations against him.

The next day he flew to Ireland with Velma, but telephoned Eddie before leaving Cambridge. He told his brother that his duty was to look after the four Pye companies of which he was a director 'to the utmost of your abilities. Remember, Pye comes first. Pye is what matters.' To Rue, who was waiting at Dublin airport, he said, 'I am ashamed of myself. Today I have signed the most dreadful document.' She asked him why he had not fought, but he gave no answer. He *had* fought, but he had lost, and all that was left was to go to Lisselan and shut its gates firm behind him.

Rue suspected Velma had something to do with the sudden capitulation. A hard drinker and hungry smoker, Velma was no longer in the best of health; she was also famous for keeping a close watch over a husband other women often found attractive. One night at the Stork Club in London she sprayed soda water over a waitress she thought was flirting too openly with C.O. At a more formal event at Grosvenor House she threw a tonic water bottle at a woman sitting at another table whom he had asked to dance. Inevitably there was gossip in the family about mistresses, and Velma may have welcomed the chance to have him to herself.

She was also better placed than most to know what had gone wrong at Pye. Some people thought that if she had been on the board the worst mistakes might have been avoided, and she may have understood that this was a fight her husband could not win. It did not mean she was reconciled to what had happened. An Irish friend who visited Lisselan that summer was struck by how C.O. 'sat hunched in his chair seeming to get smaller and smaller' while Velma delivered a violent, hour-long attack on the 'traitors' of Cambridge.

John did fight. When Pye cut short his employment in May as a first step to removing him as a director at the annual general meeting to be held later that year he at once resolved to battle it out. C.O. seemed too lost in self-pity to think about him. Peter Hoos was a friend of John and had heard much about C.O., but when he met him for the first time in 1966 he got the impression of someone who may have once been a great man, but was no longer. Certainly C.O.'s answers to the letters of sympathy he received showed little of his former spirit. He complained to one friend of having been 'forcibly removed for no reason whatsoever except that they want a change of management.' 'They' were those 'people in Cambridge . . .[who] cheated and lied in a most disgusting manner when one had spent a life-time in building them up'. A letter from Dadie Rylands of the Cambridge Arts Theatre particularly pleased him: 'I cannot help thinking you have faced what Shakespeare thought to be the worst of human vices – ingratitude.' The sentiment matched so well his mood of injured innocence that he quoted it in some of his own letters.

Only Australian friends dared tell him to 'cheer up and try to treat triumph and disaster in the manner advocated by Kipling'. His tough Australian lawyer Sir Eugene Gorman was exasperated by his descent into self-pity. 'You shouldn't need me to tell you that nothing is to be gained by worry over the past or regrets . . .over errors of judgement.' C.O. did manage a more stoic note when writing to loyal employees at Pye, emphasising over and over that 'the one thing that matters is the company'. Urged by one Pye friend to imitate Churchill and fight he answered that there were times when fighting did more harm than good, that he had to think of the interests of Pye's employees and shareholders, and 'for quite selfish reasons' the Stanley family's own holding of more than 2.7 million Pye shares. What he could not tell them was that he still hoped Philips would play the *deus ex machina* and save both something of his company and his pride.

But the demolition of the Stanleys' reputation still had a long way to go. Those at Cambridge who were once close to C.O. found themselves forced to behave like conspirators, daring to meet only briefly and talking in whispers. As for Duncan and his allies, guilt at their act of parricide and fear that C.O. might still somehow return may explain their vindictiveness. Brinkley took on the role of chief prosecutor and engaged private detectives to follow Fred Keys in the hope of finding a reason to get rid of him (they found nothing). The atmosphere of witch-hunt intensified when the Cooper Bros investigation claimed Charles Harmer as a victim. His supposed crime concerned Alec Norman, a firm that for many years supplied cars to Pye before being acquired as an investment by the Septangle Trust. Harmer was a member of Septangle and it was also his job to negotiate the terms on which Pye bought cars from Alec Norman. When the Cooper team discovered this they jumped to the conclusion that Harmer had allowed the car firm to overcharge Pye to the benefit of Septangle's members, and he was suspended from work while the accountants went deeper into the records. They found no evidence of wrong-doing, and had to apologise to him.

What the accountants did discover was more evidence, if that were needed, of the idiosyncratic way C.O. ran his company. There was no fraud or feathering of private nests at Alec Norman, but there was an odd arrangement – in this case caused by C.O.'s wish to reward some senior colleagues – that could have been abused. Cooper Bros' reports on other parts of Pye revealed many things that were shocking from the point of view of business efficiency, and sometimes inappropriate even in a less regulated age, but they found nothing criminal. Bad administration was endemic throughout the group, partly the consequence of too many subsidiaries and too little central control. There were arguably inappropriate ventures such as Pye's involvement with Manx Radio. Stock records were generally poor and almost everywhere the accounting was deplorable. Taken together it made a case for saying C.O. had to go.

Yet the intensity of the new Pye board's feud with the Stanleys and the secrecy of Cooper's work – their reports were never published, John was only allowed to see part of the enquiry into Gibbard – encouraged rumours that all sorts of unlawful activities had been going on at Pye. It was not the investigating accountants' brief to try to understand how C.O. regarded the company he had created and built up as his own, even though he no longer owned it. He acted autocratically, often with little regard for legal technicalities, in whatever he decided were Pye's best interests, seeing no difference between them and his own. As long as Pye did well no one inside the company, and precious few outside, had suggested he was doing anything wrong.

Cooper Bros' hunt for compromising material was accompanied by a fierce assault on John. Not content with dismissing him on grounds of incompetence the directors added accusations of malpractice and dishonesty, leaving John no choice but to mount a legal defence. Here C.O. did help, and on the advice of Ellis Birk put John in touch with the latter's law partner Julius Silman. By the end of their first meeting Silman was satisfied that, though his client might be guilty of incompetence or mistakes, he was 'transparently honest'. This may also have been apparent to Duncan's lawyer John Mayo. Silman told Mayo that, charges of incompetence apart, the accusations against John contained 'some of the silliest and most unworthy claims' he had seen in cases of this sort. One charge raised by Mayo was that John had stolen pencils from the Pye office. When Silman asked if he really meant to press it Mayo looked embarrassed, and said his clients insisted that this and similar accusations be made. Mayo had advised Pye against making personal attacks on John, but Duncan, Jones and Brinkley overruled him.

Silman advised John to sue Pye for defamation and libel, and the greater the intensity of the son's battle the more embarrassing the silence of the father. C.O. was still honorary president of the company whose directors were accusing John of dishonesty. And though he would say nothing to help him, he was planning to go public himself with a claim of £116 000 unpaid commission that Pye said he had waived and he insisted he had not (he promised to use the money to help employees who suffered in the aftermath of the palace revolution). His silence allowed the impression to grow that John alone was responsible for what had gone wrong at Pye, even though he could have done nothing of importance without C.O.'s approval or command. John also knew he needed more than his father's word to win his legal battle. He needed records, and C.O. had never believed in records. John now recalled with regret his father's habit of disagreeing with any figures he did not like and getting Fred Keys to change them. Records that had been rewritten, or records that did not exist at all, were poor weapons in the fight ahead of him.

Help came in September from a shareholders' committee which lobbied for John to remain on the board of Pye at least until his case against the company was heard in the courts. The committee was in fact a form of self-help, for John was its inspiration and C.O. its source of finance, though he kept this secret by having Cecil Rieck set up an apparently independent fund to pay the committee's expenses. Its members included Peter Hoos and other of John's friends, and was chaired by the film-maker Roy Boulting. John Gorst offered the services of his new public relations company and Rieck took charge of organisation.

The committee spent a considerable amount of money tracking down the little shareholders whom C.O. had always charmed and printing proxy voting forms for use at the annual general meeting. It soon discovered that by no means all C.O.'s little people were still sweet on the Stanleys. A woman who had attended the AGM after the Ekco takeover said she had not forgotten how C.O. was 'evasive and frivolous on the question of the ways and means by which such a deal was to be financed although it was . . . of grave concern to the shareholders'. Others were angry about the chaos at the company they had invested in. 'Frankly I could not care less whether Mr Stanley (one or more) is removed or shot. I *would* like to see this deplorable scrapping cease.' Another thought the whole board of Pye was at fault – 'they are incapable of running a Christmas Club'. John did find allies, though, and by late autumn the committee calculated that together with the 5 to 6 per cent shareholding of the Stanley family they had the support of 17 per cent of Pye's shareholders.

Pye's publication in October of the results for 1965–1966 brought more bad news. They showed a total of £9 million recurring and non-recurring losses, while losses caused by poor sales of radios, TVs and domestic appliances wiped out all the profits made by Telecomm and other technology-based subsidiaries. The high level of unsold stocks of TVs had forced the closure of the once-prized Ekco plant at Southend. Cooper Bros identified a loss of £2.8 million at Gibbard, less than Duncan's earlier estimate but bad enough. A pre-tax loss of £410 000 confirmed fears that no dividend would be paid. The *Financial Times* columnist Lex pointed to Pye's short-term borrowings of £20 million and wondered how the company could recover unless someone relieved it of its troubled radio and television production. Others argued the figures looked worse than necessary because Duncan had loaded them with losses from previous years that C.O. had postponed reporting. Perhaps, but another sharp comment in the *Economist* that Pye's troubles were 'just one more example of the dangers of salesmanship unrestrained by cautious accountancy' offered no comfort to the Stanleys.

With the announcement that Pye's annual general meeting and the vote on John's expulsion were to be on 17 November the battle grew ever more ill-tempered. C.O. was still nowhere to be seen in England, having taken the opportunity of Pye (Ireland)'s AGM in Dublin to give the impression that all was well in his world. A reporter from the *Irish Times* had never seen him so charming. 'The volatile, outspoken' man they all knew had been replaced by a 'demure [and] witty . . .Mr Stanley [who fell] over backwards to answer questions'. Meanwhile Duncan was writing to all Pye shareholders to put the case that John was guilty of incompetence and of making commitments without board approval, and that he bore 'a major share of responsibility' for Pye's financial difficulties and its 'unsatisfactory and loss-making ventures in general'. Duncan made no mention of the criminal charges against which John was preparing to fight (Julius Silman thought the 'refusal . . . of this rather gentle only son . . .to be walked all over' had taken Pye's directors by surprise). Duncan's letter also referred to discoveries made by Cooper Bros about Pye (Ireland) that he said were causing 'great concern'. He allowed the sums involved were 'not substantial', but made much of a suspicious-sounding 'memorandum account, known neither to the auditors nor to the board of Pye as a whole, out of which various payments have been made'.

The tone suggested Cooper Bros had uncovered some sort of slush fund. The truth was not so lurid, though still scarcely helpful to C.O. or John. Pye (Ireland) was still largely a Stanley-owned company in which Pye of Cambridge held only 20 per cent of the ordinary shares. What Cooper Bros eventually described in its full report on the Irish company was not fraudulent, but certainly was irregular. Between 1949 and 1961 Pye (Ireland) paid Pye of Cambridge a total of £64 000 for technical services. The money went into a 'suspense account' and, for reasons C.O. later explained, was not recorded in Pye's own books. A third of the sum remained unspent, and the largest single payment from the supposedly mysterious account was £9670 to Alan Bradshaw, representing his salary and pension contributions as a director of Pye (Ireland). A lesser amount was paid to John as another of its directors. C.O. authorised these and other payments without formal reference to the board of Pye even though some of the money was used to the benefit of his Irish venture rather than Cambridge. Other amounts were spent in ways that were arguably to Pye's own advantage, recipients including a Conservative Party pressure group and the Irish politician Erskine Childers, whose overeager advocacy of Pye may have harmed C.O.'s bid to set up Ireland's first commercial television company. Childers received £1950 as salary during three years in the 1950s when he was out of government (at this time C.O. also made him sales manager of a new telecommunications company started by Unidare). The 'suspense account' was more evidence, if that was needed, of C.O.'s proprietorial attitude to his companies. They were all his children, and their resources his to use as he saw best. Though not the sinister arrangement hinted at by Duncan, it was not likely to please the institutional investors on whom John's fate now depended.

Both sides continued to argue their case in letters to shareholders, and the shareholders' committee added a disingenuous message of its own. Without revealing that its purpose was to campaign on John's behalf, it proposed voting out Duncan, whose past business record the committee called 'discouraging'. John used his letter to blame his fellow directors for resisting his attempts to bring down the cost of Pye television sets. He pointed out that Brinkley, Jones and Harmer, all now joint managing directors with Duncan, had taken part in all the decisions leading up to Pye's acquisition and development of Gibbard. Of C.O., the person whose word was decisive in these and all other important matters, he could say nothing. It was like explaining the Old Testament without mentioning Jehovah.

Yet C.O. was not silent in his own defence. A week before John's fate was to be decided at the AGM he called a London press conference to justify his claim to the unpaid commission he was demanding from Pye. He also dealt with criticism that he had used Pye, a public company, to give employment to his family. He admitted that over the years ten of his relations had worked in the Pye empire (the figure included members of Velma's family in Australia) but said they were often underpaid, which was indeed true. He had the managing director of Pye (Ireland), Dillon Digby, explain the suspense account that Duncan made so much of, while Cecil Rieck took questions about Family Television, a rentals company owned by Pye executives and then, Cooper Bros alleged, bought by Pye for an inflated price. C.O. revealed nothing about his continuing hope for a deal with Philips though he had travelled to Eindhoven to pursue it, and presented himself as an elder statesman still ready to do his duty by his firm. 'If I can be of any help whatever in the management [of Pye], I am prepared to sacrifice myself to get this company . . .back to its former standard.' He said not a word about John except that he would not be defending him 'because he is quite capable of defending himself. But I won't be speaking against him.'

His performance did not win the admiration of industry colleagues such as Sir John Clark of Plessey, who thought C.O. had 'abandoned [his son] and lost his reputation in a matter of months'. There was no way of knowing what John thought about C.O.'s performance; as usual he kept his thoughts and feelings to himself. Father and son were still neither meeting nor talking on the telephone. John did know C.O. was seeing Robert Renwick, still Pye's broker and therefore Duncan's adviser too. Renwick told C.O. he hoped to persuade a couple of big institutional investors either to vote for the shareholders' committee motion or to abstain, but when John met the financial institutions in mid-November they told him he was unlikely to get their support.

Peter Hoos stayed with John and Liz in London in the last days before the AGM and found his friend exhausted and depressed. C.O., now at Lowndes Place, was still refusing to say whether he would attend the AGM or at least write a statement to be read out at it. Liz urged John to take the initiative and talk to his father. She felt John's family had abandoned them, but not even stormy arguments could persuade him to go to his father. The atmosphere only lightened in the afternoon of 16 November when Hoos got an unexpected telephone call from Roy Boulting summoning him to the home of Lord Goodman. The solicitor Arnold Goodman was a favourite adviser to the powerful, and included the then prime minister Harold Wilson among his clients. He was also chairman of Boulting's production company British Lion, and at a meeting there earlier that day had told the film-maker he was making a fool of himself by supporting such an obvious loser as John Stanley. When Boulting told him the committee had 17 per cent of Pye shareholders behind it Goodman, ever sensitive to the possibilities of power, showed interest. He thought the percentage might be big enough to force the meeting's adjournment and, sensing an interesting role for himself, asked to meet the rest of the committee.

Hoos found Boulting and Rieck at the solicitor's gloomy flat near Broadcasting House, where Goodman was getting ready to go out to dinner. They set about business, while a manservant first undressed the overweight lawyer and then helped him into his dinner jacket. Midway through this operation Goodman announced he was hungry, and the servant brought him a large cooking apple which he ate down to the core. Goodman quickly identified Renwick as the person with greatest likely influence on Duncan and, after some energetic telephoning, tracked him down to a private dinner for the new Tory leader Edward Heath. Renwick refused to come to the telephone, but Goodman sent him a second message threatening chaos at the next day's meeting if they did not talk.

When the outmanoeuvred Renwick picked up the receiver Goodman explained that as representative of the shareholders' committee he intended to demand the AGM's postponement. He added that he might also publicly accuse Renwick and Greenwells of playing a double game with Duncan and the Stanleys. The two men met late that night when the dinner for Heath was over and agreed that Goodman would talk with Duncan and his advisers before the formal opening of the meeting.

The following morning Lex wrote in the *Financial Times* that, in spite of what the Stanleys said, Pye's only hope was to get out of television production, and 'for that reason alone [share]holders should support Mr Duncan'. Hopes raised by Goodman's intervention did not last long. The lawyer and members of the shareholders' committee got to the Connaught Rooms by 10 o'clock, an hour before the AGM was due to start. Goodman vanished into a private room with Renwick and Duncan, coming out an hour later to tell the committee he 'could not save the Stanley family' and the vote on John's dismissal would have to go ahead. He did win concessions on the appointment of independent outside directors, which was another of the committee's demands, but it was small consolation for John's friends, who by then knew that C.O. planned to spend the morning at Lowndes Place, emerging only after John's execution had taken place to have lunch with Robert Renwick.

John arrived at the Connaught Rooms with Silman knowing that his father would neither be present nor (possibly on Renwick's advice that it would damage Pye further) take this last chance to say something on his behalf. Silman thought his 'very shy, hesitant' client showed 'extraordinary dignity from start to finish'. He believed John accepted that someone had to take the blame for Pye's troubles, and had only fought back because of the malice of his former colleagues' attack on his integrity.

The thousand people packed into the Connaught Rooms slow hand-clapped when Goodman's behind-the-scenes negotiations delayed the 11 o'clock opening. Accompanied by Mayo and David Hobson of Cooper Bros, Duncan led the Pye directors into the hall to a mix of cheers and boos. John and Fred Keys had a similarly ambiguous reception when they took their seats at the end of the line, Keys upset at not being allowed to take his usual company secretary's place at the chairman's right hand. Duncan struggled to control an audience whose restlessness increased when the microphones failed. There was barracking from small investors who supported John and from some Pye employees who had come from Cambridge, among them Queenie Culverhouse, Pearl's assistant in the wartime system of outworkers.

Duncan barely managed a sentence before a woman called out to ask who he was. John Mayo smiled; Pye's new chairman did not. Goodman got up to explain his role, and spoke too long. During legal duelling over whether it was proper to ask questions about Pye (Ireland) an intervention by Ellis Birk, representing C.O., caused a flicker of excitement that Pye's founder might be sitting at the back of the hall. When Duncan asked for the accounts to be accepted on a show of hands there were calls for a poll, but they died away when he announced that a count of proxy votes showed 13 million for and only 5 million against.

John no longer had any hope when he got up to speak on the motion opposing his re-election to the board proposed by Duncan and seconded by Rupert Jones, but he lost his composure only once. When he complained that Pye was treating him as an embarrassment a voice shouted from the hall, 'What do you want – a knighthood?', and John snapped back 'belt up'. He recovered himself to ask why he should bear the

Figure 10.6   John Stanley, arriving at the Connaught Rooms to face his downfall alone at the shareholders meeting, 1966

blame for decisions others sitting on the platform with him had shared in. It was a good point, and explained why he so embarrassed his former colleagues, but it was not likely to impress the institutions which, whatever Renwick may have promised C.O., were all preparing to vote against him. At the end John said he knew his head was on the block. The City staff of the *Times* were surprised that a man who had just

Figure 10.7   *The final moments of the Stanley dynasty at Pye as the shareholders vote to sack John Stanley from the Board of the Company*

made a fine speech should mar it by such a 'defeatist' remark. Had they known the whole of the story they might have wondered that he managed to make a speech at all. A show of hands was taken and only a scattering went up in John's support.

When the meeting ended Duncan, Harmer, Brinkley and Jones kept away from John as though he had the plague. Brinkley, his protégé at Telecomm, cut him a second time as they were leaving the building. The *Daily Mirror's* double-page report of the morning's events had a photograph of John walking impassively away from the Connaught Rooms. It also carried a short statement from C.O. He had made 'a great sacrifice', he said, by not going to the meeting. It was nevertheless his duty to 'stand aside and avoid adding anything which, although telling what had happened between 30 March and 7 May this year, could at this time only do more harm to the company'. His choice of dates suggested he was still more sorry for himself than for his son, for they encompassed only the crisis that led to his own banishment from Pye.

E.K. Cole died the next day, an apt coincidence because if any one thing sent Pye into decline it was C.O.'s decision to merge with Ekco. Dennis Fuller, most clear-minded of his admirers, said 'he single-handedly destroyed the one thing that he spent most of his life building'. He also came close to destroying his only child.

Sources

COS files: 2/1 (John Stanley's December 1966 history of Gibbard), 2/12, 2/13 (C.O. Stanley's negotiations with Philips), 2/21 (John Stanley's notes on fight against Duncan), 2/15, 2/5, 2/9, 4/2.
COS/JOS2/1/1 (shareholders' committee), COS/JOS2/1/2, COS/TWEM1/1, COS/TWEM2/1–2.
Simoco boxes: Box 1 (Cooper Bros reports).
A–Z files: Fritz Philips (Pye and Philips).
Pye main board minutes 1965–1966 in COS/TWEM2/2.
Interviews: Ellis Birk, Don Delanoy, Sir John Clark, Dennis Fuller, Peter Hoos, Fred Keys, Julius Silman. Peter Threlfall, Michael Worsley.
Michael Bell interviews: David Hobson, J.B.H. Jackson, John Stanley.

Chapter 11

Son and father

After the ordeal of the Connaught Rooms Velma and Peter Hoos took John and Liz to a restaurant across the street in an effort to keep up their spirits. C.O., preoccupied with his own hurt feelings, made no such gesture towards his son. He knew already that some old friends and colleagues now found him an embarrassment. Renwick had told him in the summer that fellow directors of British Relay Wireless (BRW), a business the two of them virtually brought into being, wanted to be rid of him. Renwick was 'very upset,' but advised it 'could be best for BRW' if C.O. did not put his name forward for re-election to the board. That autumn C.O. had accused Sir Richard Powell of trying to get him off the platform at the Institute of Directors' annual conference. He told Powell he should have protected him 'from the attacks of unscrupulous directors who wished to destroy me'. Powell replied that the 'fierce controversy' surrounding C.O. made it sensible for him to 'skip' the meeting.

Father and son at least had the satisfaction of seeing Philips take over Pye. Eindhoven's first attempt in November 1966, during which they acquired a large part of the Stanley family holdings, was outbid by Thorn, but a new offer two months later was successful. The takeover of a leading British electronics company by a European group required discussion in Cabinet (the government had the power to use exchange controls to block the bid). Ministers considered Pye's telecommunication and instrument businesses 'technologically interesting, expanding and profitable', but thought little of its prospects in radio and television, a judgement that proved correct, though C.O. would have furiously contested it. The Philips' offer was allowed to go ahead.

Some saw the Dutch victory as vindication of the Stanleys. When the discovery of a £1 million stock overvaluation forced John Brinkley out of Telecomm, and Charles Harmer and Rupert Jones lost their managing directorships, the press wrote of a 'Stanley family ... morally restored in influence though not in power'. C.O. was certainly active in advising Philips. He wrote assessments of senior executives at Cambridge for Engels, and in February travelled on Philips' behalf to look at Pye's offshoots in Australia. He also tried to persuade Eindhoven to pick the successful managing director of Unidare, Percy Greer, as Pye's new managing director. Philips ignored his advice and appointed Peter Threlfall, a rebuff that reminded C.O. that an honorary president had more prestige than power. Perhaps he had reason to expect more. At the height of the takeover battle Jan Engels had summoned Cecil Rieck and told him that if Philips won they would 're-instate C.O. Stanley as the effective head of the business'. If Engels said that, he can hardly have meant it. There was no way the Philips style of methodical management could be reconciled with C.O.'s old formula of inspiration and autocracy.

His lack of real power under the Dutch regime was soon made plain at Arks, C.O.'s first business venture that survived on publicity work for the Pye group. In 1964 C.O. had appointed his former personal assistant Anne Butler as Arks' managing director, an unexpected promotion that fed gossip about the nature of their relationship. Butler was tough and efficient in C.O.'s office, but she had little experience of advertising. Her work was made no easier when C.O. appointed Velma and Rue to the Arks board, and her prospects collapsed entirely when Philips made plain it had no interest in supporting the company. She died of a heart attack shortly before Arks was taken over by an American agency in 1967.

C.O. was left with ATV as his only British interest, but, with its success established, his contribution was necessarily less than in the challenging early days. (When he eventually retired from the ATV board in 1975 he enquired about a directorship in one of its smaller subsidiaries, but nothing was offered him.) Quarrels with the Inland Revenue, and with accountants he thought insufficiently pugnacious, added to his disillusion with Britain. He told friends that if Velma had not been unwell, and so fond of Sainsfoins, he would have sold the house and been 'out of the country so quickly it is nobody's business'.

There was perhaps another reason for his readiness to leave. As long as he was in Britain he would fret over how he lost control of Pye, and look for further proof of treachery. He was obsessed by the suspicion that Renwick had played a double game, and even four years after the palace revolution presented him with what he took to be conclusive evidence. This was a 'signed personal affidavit from a man of great integrity, well known to you', according to which Renwick promised Frank Duncan he would help get him a seat on the Pye board. Renwick may not have answered, knowing it was pointless to try to pacify C.O. in full vindictive flight.

Without even his father's dwindling business interests to occupy him, John withdrew into himself, leaving Liz to protect him from what seemed a hostile and unfeeling world. He knew that many thought his dismissal from the board of Pye implied dishonesty as well as incompetence, and as long as his claim against Pye remained unsettled he could not interpret Philips' changes at Cambridge as any restoration of his 'moral influence'. Peter Threlfall's agreement to become Pye's new managing director upset him too, for he could not understand how someone he thought a loyal friend could accept the job before Pye had apologised to him.

The change in John's manner and appearance shocked those who knew him. A cousin found him 'a shadow of his former self'; Peter Hoos thought he was 'absolutely shattered'. The future seemed even bleaker when he was found to have dangerously high blood pressure. John decided he could not look for another job until Cambridge had satisfied his demand for compensation, and the *Financial Times* ran a story about Pye's former deputy managing director passing the time by painting toy soldiers. It was true. He bought the lead models at a shop in London and took them back to his small estate outside Newmarket where he added the colours. He no longer took the family for Sunday lunch at Sainsfoins, and dealt with the pain of his relationship with C.O. by wrapping it in silence. Neither then nor later did anyone hear him say a critical word about his father. The depth of his bitterness may be judged by one rare remark some years later to a newspaper that 'everybody who could have given me a job or helped me passed on the other side'.

Summer holidays at Mill Cove eventually brought a re-establishment of relations with C.O. Velma still drove him over from Lisselan to eat their separate, and superior, Sunday lunches at the edge of the swimming pool and later C.O., always alone, rowed far out to sea in his dinghy. After John underwent stomach surgery C.O. took the whole family to Jamaica where he and Velma stayed in their favourite grand hotel but put John, Liz and the children in a cheaper place along the beach.

The relationship eased further when both men settled their claims against Pye, though the battle left John with a lasting suspicion that Philips had not played fair with him. If Rieck's account of his December 1966 conversation with Engels was accurate, the Dutch agreed that if they won the takeover battle, 'injustices that had occurred would be looked into and corrected'. In fact C.O. had to fight three more years for the £100 000 unpaid commission that he had pledged to use to help 'hardship cases' such as Rieck and Norman Twemlow. He got it only after threatening to resign as the company's honorary president and take Pye to court.

What was almost a game for the father was of utter seriousness for the son. At the end of 1968, after two years of fruitless negotiations,

Figure 11.1   C.O. Stanley inspects a book of employees' signatures presented to him as honorary President of the Pye group of companies, 1966

John had a long meeting with Jan Engels. The latter professed to have no exact knowledge of Pye's charges against John. He also seemed unaware that John was asking for restoration of his pension rights and share options, for legal costs, and for an admission that he was 'the victim, not the criminal'. John told Engels he was ready to appeal up to the House of Lords if necessary to get access to the Cooper Bros report, and pointed out that if the matter came to court his action would be as much against Philips as Frank Duncan, who had just relinquished the chairmanship of Pye.

The thought of Philips being dragged into the dispute alarmed Engels, who countered by suggesting that going to court might also implicate C.O. John said he had 'thought about that' and decided that 'a full exposition would recover some of [C.O.'s] reputation rather than otherwise'. That was hard to believe, but John may have believed it, for it was harder still to imagine him doing anything he thought might damage his father.

Some months later Pye's new chairman, the former Conservative Chancellor Lord Thorneycroft, telephoned Silman's partner Ellis Birk and said, 'John's position is old history. The board has all changed. It's high time the matter was sorted out.' In October 1969 Pye met most of

John's financial demands, and also issued a public statement 'to make it clear that at no time [had] they doubted the honour and integrity of Mr Stanley'. Silman guessed that Thorneycroft was relieved to settle the matter so easily and urged his client to hold out for more, but John had had enough and the lawyer did not press him. He knew John had got what he wanted most: Pye's retraction of any suggestion that he had behaved improperly.

The victory came not a moment too soon. In early 1968 John and Liz at last managed to sell their house at Newmarket and moved into a one-bedroom flat in Primrose Hill, where their four growing sons had to sleep on fold-up beds in the dining room. It was a bleak period for the whole family, but above all for Liz, who at times felt alone and without comfort. It was a principle for John that capital, in this case from the sale of his Pye shares to Philips, should be used only to finance future businesses, and this may have persuaded him to rent such a small flat while they looked for a London house to buy. The choice may also have reflected a self-punishing depression. His fragile state of mind can be judged by the tears this most reticent of men shed when he discovered that the model German train C.O. gave him when he was a boy had been lost during the move to Primrose Hill.

The melancholy of C.O.'s situation was plain when Velma died in 1970. Chronic coughing and even a bad bout of pleurisy had not persuaded her to give up the cigarettes her maid Mary Sullivan smuggled in for her, and which she smoked, like a guilty schoolchild, standing at an open window. C.O. could not settle after her death, and within a year married Lorna Sheppard, which was what Velma had often told him he would have to do. The two women had been at medical school in Australia together and Lorna went on to become an authority on addiction. She had money of her own and matched Velma in elegance. An Irish friend thought C.O. looked up to her because of her intelligence. 'For him she was like a good mahogany table that other people would admire, except that she had two legs rather than four.'

C.O.'s siblings did not welcome the latest wife to rob them of their brother. They found Lorna quite as domineering as Velma, and even more aloof. Irish friends sensed she did not like their country and had difficulty understanding their accent, but with a new wife to organise his social affairs C.O. was able to resume something like his old life. Shocking evidence that he was in danger of being forgotten came when Anthony Sampson's 1971 *New Anatomy of Britain* described him as 'now dead'. C.O. was furious, wanted to go to law, but settled for the insertion at the offending page of a slip that read, 'Mr C.O. Stanley ... is very much alive and active'. Active he was, but increasingly only in Ireland. After the palace revolution Duncan had replaced him as chairman of Unidare, but he could not touch the management of Pye

(Ireland) in which Cambridge only had a minority stake. This was the last surviving piece of C.O. Stanley's electronics empire, but even by his own account it was a meagre fragment. With capital employed of less than £1 million it ranked 40th in the list of Ireland's industrial companies and had only 350 employees. 1966 had been a bad year, with a record loss that C.O. blamed partly on the reduction of tariff barriers by an Irish government intent on joining the European Community. 'Every time an import duty is reduced', he complained, 'Irish manufacturing industry becomes a less attractive proposition.' The truth was that Pye (Ireland) suffered from many of the same flaws that brought C.O. low at Cambridge. Credit restrictions destabilised the television market, dealers needed expensive financing that the company could ill afford, and there was increasing competition from foreign imports. C.O. ordered cutbacks and, as before in England, put his faith in the arrival of colour television, which he predicted would bring a 5-year boom.

He was wrong. The Irish government paid no attention to his warnings against 'low-cost mass-produced imports' and by the mid-1970s Pye (Ireland) was floundering. At the end of 1974 its accountants wrote a report that sounded familiar to anyone who knew what had happened at Cambridge. Accounting and stock management were poor; there was a large undiscovered liability for the late payment of VAT and PAYE; hire purchase debts were out of control. The company owed £750 000 to Pye and Mullard, who were its main suppliers, and £800 000 to the banks that financed its dealers' hire purchase operations.

C.O. and Lorna were in Australia in early 1975 when he learned that Peter Threlfall, worried about the stability of a company in which Cambridge owned a share, had sent a pair of hatchet men to Dublin. Within weeks the Bank of America called in its loan of £240 000, and Pye (Ireland) survived only by submitting to a reconstruction that brought a new management and C.O.'s resignation as chairman. Although not responsible for day-to-day management he never gave any sign he thought the company was badly run, and in his remarks to the 1974 annual general meeting put the blame for its difficulties on unforeseeable 'hazards' and the usual suspect, unfair foreign competition.

A year before the climax of the palace revolution Pye Telecomm had launched the Pocketfone, Britain's first more-or-less pocket-sized radio telephone. It had a range of only 5 miles and was clumsy by the standards of the early 21st century, but it allowed an enthusiastic press to predict the day when anyone would be able to call home from the top of a bus. It was perhaps inevitable that John would be drawn back into the world of mobile communications in which he had spent his earliest and happiest days at Pye. His return marked the beginning of his ascendancy over his father, albeit an ascendancy he never claimed and C.O. never recognised. John had the right combination of technical imagination and the will to fight the officialdom that still controlled Britain's airwaves. And he had in a company called AirCall a suitable vehicle to pursue his passion for mobile communications. AirCall's origins went back to 1956 when he set up a private company, Telephone Answering Services (TAS), after a visit to the United States. His plan then was to develop an answering business of the kind he had just seen in America (the only company already offering this service in Britain was American-owned). C.O. showed no interest in the project and kept Pye out of it though, to do him justice, few at the time apart from John understood what the venture was about.

By 1960 TAS was operating from an office in Wardour Street and providing an answering service in London for doctors, restaurants and actors. In many ways it was a typical Stanley operation. The directors included his aunt Rue and an old school friend, Michael Worsley. It was also, typically, strapped for cash. When Rue started up TAS operations in Birmingham she equipped its office with furniture from the bedroom of her son Warren who was away at university. Later both Warren and Rue's daughter Sally joined the company.

With Pye Telecomm leading the way in British mobile radio for taxis and other businesses it was natural for John to want to link mobile radio with a telephone answering service, and in 1960 he wrung from an ever obstructive Post Office the first frequency for two-way message handling. The following year TAS launched AirCall, Britain's first radio answering service. John quickly pushed its development beyond London, and in 1965 signed a 20-year agreement with the British Medical Association to provide a radio service for GPs. The idea was controversial and the government at first hostile, but the advantages of such a service were soon obvious. Using AirCall's system the 150 GPs of Sheffield could cover all emergency night-time work with just three doctors on call. The BMA Emergency Treatment Service eventually covered two-thirds of Britain and its success served John as a weapon in his future battles with the Post Office.

TAS and AirCall were little more than a hobby as long as John was Pye's deputy managing director wrestling with the group's mounting problems. This, together with insouciance about financial discipline perhaps learned from his father, almost brought the new ventures down. AirCall grew too fast. When the 25-year-old Warren Tayler joined it in 1965 John made him financial controller and gave him a cheque for £40 000 to keep the business going. When Tayler tried to pay the wages with it he found the money had already been spent.

As soon as John started to recover from his expulsion from Pye he bought a company that made equipment for the food packaging industry. He made other small investments too, but he was soon giving most of his time to AirCall. The memory of 1966 affected him in two ways. He ensured financial prudence by keeping Cecil Rieck at his side (Warren Tayler thought Rieck a Jonah who always erred on the side of pessimism), and he resolved never to let the control of any new venture slip out of his hands. He would neither give pretext for another Cooper Bros investigation, nor allow anyone the chance to launch a second palace revolution.

John knew that the arrival in Britain of 625-line television left the old 405-line channels free for mobile radio, and in 1970 won for AirCall the first licence for a radio paging service using digital signalling. Eventually the privatised British Telecom followed with a similar service of its own. He also set up AirCall (Teletext) to provide text messages on the under-used space on television channels but both ITV and the BBC were obstructive, and it was some time before the enthusiasm of the financial institutions made Teletext a profitable venture.

It seemed to Warren Tayler that his cousin had a new idea every day, though some of them were hard to put into practice. John was still not always easy to work with. Shyness could make him seem intimidating, and, when lost in thought, he might walk past people without noticing them. C.O. led by attraction and forcefulness. John won loyalty by his enthusiasm and originality of mind. AirCall entered the computerisation of communications (it was the first British company to import an Apple computer from America) and also started to operate abroad. Expansion, though, was held back because of lack of capital, the consequence of John's insistence that he keep a majority shareholding at all times.

The contrast between John's brave new world and his father's old and diminishing one might have been painful to C.O. if he had noticed it, but he gave no sign that he did. He had never taken much interest in his son's separate activities, though Pye did supply AirCall equipment on preferential terms. Later C.O. helped AirCall ride out cash problems with at least one loan of £30 000, and did not object when family trusts advanced larger amounts, but he never bothered to visit AirCall's offices. There is no record of him speaking with pride of what John was doing.

He had problems of his own. Much as he respected Lorna and needed her companionship, she was not another Velma. She was often busy with her own work. She was relentlessly smart, and given to wearing expensive jewellery and elaborate hairstyles that were out of place at Sainsfoins, let alone Ireland. C.O.'s servants still thought of Velma as 'our Mrs Stanley', and Lorna further disturbed domestic harmony by bringing her own maid and confidante, a refugee from the Czech Sudetenland who tried to wean C.O. off Irish cooking and onto Central European food. Lorna also caused a change in C.O.'s summer habits. He had never liked driving himself, and because Lorna was a poor driver he needed somewhere closer than Mill Cove for his summer swimming

MONEY
PAGE

14, FINSBURY CIRCUS, E.C.2.
Tel. 606-1234 STD Code 01

Edited by DEREK PORTER

BLEEP! EX-PYE BOSS MAKES COME-BACK

MR. JOHN STANLEY, the former deputy managing director of Pye of Cambridge and son of its founder, Mr. Charles Orr Stanley, is bleeping his way to a second fortune and plans to make a City comeback.

The Stanleys were the centre of one of the City's most bitter company rows in the mid-60s with John being ousted from the board room and his father resigning.

SERVICES

One of its most interesting services is personal paging. For a few pounds a week a top man can be kept in touch with his office by carrying a small receiver with him.

His secretary just dials Air Call, gives any message to the centre.

Figure 11.2   Evening News, 10 June 1974

and boating. He found an austere and isolated bungalow on a cliff edge at Simon's Cove. Not many 70 year olds would have picked such a place. Driving down to it was difficult, getting out harder still. There was a small beach, but you had to scramble over rocks to reach it, which C.O. did not mind because he bathed from the rocks, and even managed to keep a small boat and lobster pots. Lorna did mind, and asked C.O.'s doctor to warn him against what she thought his dangerous habit of swimming in all weathers. The doctor refused, saying C.O. had done it all his life.

In the spring of 1974 Lorna fell on steps at Lisselan and never quite recovered. C.O. was now deeply involved in the affairs of Sunbeam Wolsey, the Cork textile firm of which he had been a director for 40 years, and chairman since 1971. Though a far more substantial undertaking than Pye (Ireland), Sunbeam also faced the threat of losing a long protected market. A series of bad results persuaded some of the directors it was time to give up but C.O., backed by the new managing director, Tom Scott, would not hear of it. The Irish press said C.O. was too old for such a difficult job, but he set about restoring profitability by closing down loss-making operations, much to the alarm of an Irish government fearful of unemployment. In true C.O. style the textile firm was also running a large overdraft with the Munster and Leinster Bank which, when it learned of Sunbeam's troubles from one of C.O.'s own directors, called in the debt. C.O. filibustered while ordering Sunbeam's subsidiaries to sell everything they could at give-away prices, and within a month he paid off the bank. A year later the Munster and Leinster was again treating him as an honoured customer, and journalists praised him for having brought off a 'splendid turnabout'. 'He had such a brain', thought Scott. 'He could see his way through situations I couldn't. He was always bloody-minded and "opposite", but you could only admire the way he worked.'

In early 1975 C.O. was in Australia with an increasingly disturbed Lorna when Scott wrote to him about an impending showdown with the unions. 'We are missing you at the helm [but] as you would say "it is now or never."' When workers supported by the Provisional IRA occupied Scott's office he wrote jokingly to C.O., 'Having a wonderful time. Wish you were here.' He meant it, too, because he knew C.O. had more 'backbone and courage' than anyone else in the firm. A rumour went round that the IRA threatened to burn down Lisselan and told C.O. he could not count on the Clonakilty fire service. There would have been little local support for any attack on his property, but he did build a 10 000-gallon water tank near the house.

Sunbeam was his Last Hurrah. He was ill when the company held its AGM in 1976 and John had to take the chair for him. Early the following year Lorna insisted against his wishes on travelling to Australia accompanied only by her unpopular German maid. C.O. went to Ireland and tried to occupy himself with Sunbeam's affairs, but after two days became disoriented and almost collapsed. Lorna herself fell ill soon after her arrival in Melbourne, and C.O. needed the help of his private secretary to fly to Australia in the hope of bringing her home. The strain was too much for him and he had to be sent back to London. Lorna died soon after and the following year, 1978, C.O. resigned as Sunbeam's chairman though Scott begged him to stay on. He could still seem his old self when he had people to talk to, but he was increasingly forgetful. Alzheimer's Disease had taken him under its slow siege.

John, who had himself gone to Australia to help manage Lorna's last difficult weeks, was soon supervising more and more of his father's life. Without a wife to support him C.O. sold Sainsfoins and moved permanently to Lisselan. John went twice a month to see him, but his thoughtfulness earned him no credit. The visits were seldom easy, and there had to be frequent discussions with lawyers and accountants on ways to limit tax on C.O.'s estate. Ireland at that time had a wealth tax and it was thought prudent for C.O. to make a gift of Lisselan to John, who hoped to live in the house when he retired. C.O. agreed,

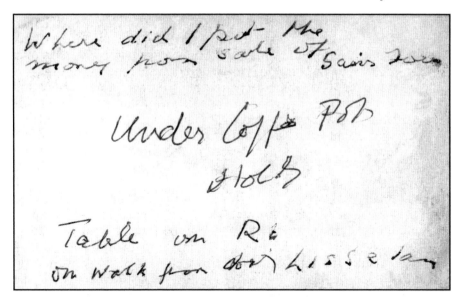

Figure 11.3   Question and answer, both written down by C.O. Stanley, as Alzheimer's Disease takes its hold of his memory, c.1980

but was never comfortable with the decision, and complained that the accountants, whose long debates now muddled him, were behaving as though he were already dead. He sometimes greeted John with a bad-tempered 'why are you here?' or even an accusation that he had stolen his money. On one occasion he ordered him to leave the house.

An Irish adviser tried to reassure the old man. 'You are the person who has to be satisfied [by the financial arrangements], and if you do not get peace and enjoyment from the wealth you have created then it was not worth creating in the first place', but there was little real peace in his life. When his London lawyer Michael Rose visited Lisselan in 1979 C.O. had trouble recognising him. He was better the following morning, getting up early as usual to make his own tea, then offering Rose a breakfast of Cornflakes with orange juice poured over them. After talking business in the morning he took his visitor for a drive in the old Ford Escort that he bought when he settled at Lisselan. 'All you can see is mine', he announced, though he had already given it to John. When he turned to head home he had to ask Rose which way to go. Later he put stickers on the dashboard to remind him where the lights and indicator switches were. Yet he still managed to drive to Simon's Cove to swim, and was 80 before he gave up setting his lobster pots.

People noticed a bleakness about him. When members of the family came to see him, and he seemed not to recognise them, some suspected

it was not forgetfulness but an old man's mischievous pretence. Relations with his sisters had often been stormy, but now the quarrels were
different. Pan went to keep him company one Christmas but after a few
days he told her to get out. She never stayed at Lisselan again. He liked
to quote to visitors Polonius' advice to Hamlet, 'The friends thou hast,
and their adoption tried,/ Grapple them to thy soul with hoops of steel',
and as he grew more forgetful he might repeat it several times in half
an hour. He had no one close to him except his son, and he kept him at
arm's length. None of this deterred John, and when his father needed
constant attention he refused to let him go into a nursing home, and
organised round-the-clock nursing at Lisselan.

The strain on John was all the greater because his own business was
now growing rapidly. Needing more money to compete with British
Telecom, AirCall became in 1980 the first company licensed by the
new Unlisted Securities Market. Unprofitable parts of the business such
as Teletext and various foreign ventures were shunted into a separate
company, while AirCall plc offered a 25 per cent stake in its British
communications and medical services worth £2.5 million. Two years
later a further offer of 24 per cent (which still left control in John's
hands) raised another £5 million thanks to the steep rise in AirCall's
share price.

The money allowed John to enter the new world of telex re-filing
using computers, a much needed alternative to the slow and expensive
telex that was still the chief means of transmitting information round
the world. The purchase of an American company that owned banks of
computers in the United States, Europe and Australia allowed AirCall
to develop something like an early international e-mail network.

The pressures of a rapidly changing business and family responsibilities (the care of C.O. apart, John gave much time to advising other
members of the Stanley clan) may have been the trigger for the first
of several strokes, but he survived them to launch his most ambitious
project yet. In 1982 he became intrigued by experiments AT&T was
conducting in the 800 and 900 MHz band. This very high frequency had
only been used by military radio links, and John guessed the American
company was trying out radio transmissions over such short distances
that they had to be using interlinking cells. John persuaded the Post
Office to give him 50 frequencies for trials with a new type of cellular
radio, but though the American Federal Communications Commission
had already granted licenses for cellular radio in cities he could not
persuade any British company to make the equipment he needed. He
decided his best chance was to persuade the government to give him a
large number of permanent channels which he could then use to attract
American backing.

He got more than he bargained for. Kenneth Baker, the minister with
responsibility for technology, was so enthusiastic he agreed to allocate

500 channels, but to John's fury ruled that half would go to the newly privatised BT and the rest be auctioned off to the highest bidder. John protested in vain to Prime Minister Margaret Thatcher, his hero since she was leader of the opposition, but in the end accepted that AirCall had no choice but to battle it out with much bigger competitors.

He entered the fight with the same confidence C.O. showed in all his campaigns, setting up a new company called Cellular Radio in which finance houses and some other small operators joined. When Racal emerged as the most threatening competitor, John protested, without success, that it was a manufacturer and not an operator, as the terms of the auction laid down. He would not admit it, but he never stood a chance. Racal had long experience of military contracts, work that John had lost interest in at Pye. Racal's relations with Whitehall were good, whereas John displayed all his father's contempt for the men in the ministries on whom the fortunes of his company so largely depended. Gerry Whent, chairman of Racal Radio, knew John had no hope of winning and, recognising AirCall as the project's pioneer, offered John a share in what he was certain would be his own winning bid. He proposed AirCall merge with Racal in exchange for 15 per cent of Racal Vodaphone. John turned him down. He did not trust Racal's offer of partnership, and C.O. would surely have applauded his son's defiance. Racal won and Vodaphone took off into the new and immensely profitable world of truly mobile telephones. AirCall's shares slumped.

In the late summer of 1984 John threw a party at Lisselan to celebrate Rue's golden wedding. There were over 200 guests and four generations of Stanleys including Eddie, Pan and Ginger, and a chartered plane for those coming from England. An Irish newspaper declared the champagne lunch in a marquee the equal of anything offered to president Ronald Reagan, who was then on a visit to Ireland. It was natural for John to want to thank the aunt who had done so much for him; some wondered, though, if the party was also a shy, and unrecognised, offer of thanks to the confused old man who went among the guests like a ghost.

The next year John suffered a stroke while on holiday with Liz in France, and died soon after being brought back to London. C.O. was not told until after the funeral, and he may never have read the obituaries that remembered his son as a prolific innovator in the fast changing world of mobile communications. By creating AirCall John had passed the only test of worth his father recognised, yet C.O. hardly seemed to notice.

Liz took over the supervision of C.O.'s care, making the regular journeys to Lisselan where the Irish staff took the increasingly senile old man under their protection. He spent much of the time in the library in front of an old Pye television set, a newspaper on his knee and

Figure 11.4   C.O. Stanley with his nurse at Lisselan, 21 June 1987

spectacles askew. The gardener took him for a morning walk; after lunch the cook came and talked to him and he sat and listened. He could still lose his temper and reduce his bedroom to chaos if he could not find his Trinity College tie.

Eventually they moved his bed into the drawing room and he died there on the morning of 18 January 1989, falling quietly asleep after talking about what he wanted for breakfast. John's last act of filial piety had been to arrange the funeral service in advance, and C.O. was buried in his best suit and Trinity tie next to Velma in the graveyard of All Saints, Kilmalooda. When the family arrived at the little country church they found a press of silent, simply dressed men and women round the door. The local people had come to mourn; 'he was loved in Ireland', Lisselan's housekeeper, Sarah Lane, said, 'because he was a fair man'.

After the service they went back to Lisselan where Liz had brought up wine from C.O.'s cellar. The wake would have pleased him. There were nuns from the local convent whom he had helped for many years, Protestant clergymen and Catholic priests, and the Clonakilty solicitor who was a cousin of the Irish revolutionary Michael Collins. Ireland reclaimed him.

Sources

COS files: 1/2, 1/2/1, 1/4, 1/8/1, 2/12–13, 2/15, 2/21 2/9, 3/4/2 (Sunbeam), 5/1, 5/5, 7/8/1 (Pye Ireland).
COS/JOS1/2, COS/LKS1/1.
Public Records Office: CAB 128/127.
Interviews: Sally Emerson, Brian Beer, Rethna Flaxman, Nuala Hall, Peter Hoos, Sarah Lane, Anthony Lucas, Ted Murphy, John O'Connell, Anne Powell, Michael Rose, Tom Scott, Julius Silman, Nicholas Stanley, Warren Tayler, Daphne Whitmore.
Michael Bell interview: Dillon Digby, Tom Linnane.
A–Z files: Dillon Digby (Pye (Ireland) annual general meetings), Sir John Gorst, Warren Tayler (John Stanley and TAS).

Source materials

The letters, documents and other material on which this book is based are held at the National Museum of Photography, Film and Television, Bradford.

The COS files

These are organised into eight groups:

COS1 – Stanley's personal papers (19 files).

COS1/1/7, the Stanleys, Cappoquin and the Rowing Club.

COS1/3, dealings with Barclays Bank from 1945.

COS1/8/1, personal documents and notes on the family.

COS1/8/2, material on the last years of Stanley's life, and also Stanley's notes on wartime work.

COS1/9, family papers from the 19th century.

COS2 – Stanley's Pye papers (23 files).

COS2/1, letters and memos covering 1960–1966. Among the most interesting are: Cecil Rieck's 1965 study, 'Angles on domestic radio and TV'; letters from R.M.A. Jones on Pye's mounting difficulties in the 1960s; Stanley's resignation.

COS 2/2 includes notes on television transmission in the 1950s, including the problems of B.J. Edwards.

COS 2/3, relations with wartime ministries, especially military radio; General F.O. Pile's letter on Pye's first attempt at a radio fuze.

COS2/5, disputes with Ministry of Supply 1950–1954 over military radio.

COS2/6, Pye's 1954 award for wartime achievements.

COS2/12, Pye (Ireland) 1956–1976.

COS2/13, correspondence with Philips 1966–1970 including J.P. Engels on Stanley's 1965 request for help and subsequent developments.

COS2/17, Pye annual reports 1937–1977. After the war, Stanley used these to ride his hobby-horses and attack politicians and bureaucrats.

COS2/19, Pye's 1960 takeover of Telephone Manufacturing Company.

COS2/21, papers on the 1966–1969 departure from Pye of C.O. and J.O. Stanley.

COS3 – Stanley's business papers other than Pye (11 files).

COS 3/1, Arks Publicity 1923–1967, including board minutes 1923–1930.

COS3/2, ATV 1960–1970. Correspondence with Norman Collins and other ATV figures, and concerning the Pilkington Committee.

COS3/3/1–2, Institute of Directors 1960–1962, Stanley's part in creating and running Export Action Now.

COS4 – Stanley's business contacts (five files).

COS4/1, correspondence with Hugh Cudlipp 1958–1965 on ATV matters.

COS4/5, miscellaneous correspondence 1929–1978. Stanley's 1929 letter assessing his own abilities and future.

COS5 – personal contacts (six files).

COS5/1, Francis Coulter. Includes letters on Stanley's honorary degree from Trinity College, Dublin.

COS6 – speeches (one file).

COS6/1, speeches and statements 1945–1964. Includes 1949 and 1950 speeches on need to end BBC television monopoly, also a copy of Robert Watson-Watt's 31.8.45 speech on radar.

COS7 – trusts and foundations (18 files). These mainly concern the Stanley Foundation and other charitable trusts set up by Stanley. However, COS7/1/6 contains notes of Michael Bell's 1972 interview with Stanley's sister Mrs Tayler (Rue) and of a discussion between J.O. Stanley, Norman Twemlow and Bell. There is also a copy of Stanley's December 1958 address to Pye executives on the crisis caused by the failure of the VT14 television set, and notes by J.O. Stanley on Pye's wartime activities.

COS/JOS – papers of C.O. and J.O. Stanley (eight files).

COS/JOS2/1–2, activities of 1966 shareholders' committee in defence of John Stanley.

COS/JOS2/3, Stanley family's Pye shareholding; John Stanley's 10.11.66 letter to shareholders.

COS/LKS – Lorna Stanley 1974–1978 (one file).

COS/VDS – Velma Stanley (five files).

COS/VDS1/1, Velma Stanley's letters 1932–1949, including 1932–1937 letters from Stanley describing his battle to win control of Pye.

COS/VDS1/3, Velma Stanley's feud with Cambridge ARP 1940–1941.

COS/VDS1/5, 1967 dispute with the Bishop of Cork over Irish land.

COS/RT – Reg Thompson's photos of Pye products (one file).

COS/TWEM – papers of Norman Twemlow, Pye director and Stanley supporter (five files).

COS/TWEM1/1, Gibbard 1959–1966. Chronicles Gibbard's problems and Pye's attempt at rescue.

COS/TWEM2/1, Pye 1957–1966. Notes on Septangle Trust. Draft Pye memorandum to Pilkington Committee 1961. Various on colour television, Pye 1965 reorganisation, and production figures.

COS/TWEM2/2, main board minutes 1958–1966. Includes 1961 Stanley correspondence on purchase tax and hire purchase restrictions, collapse in radio and television sales. Stanley's 20.11.63 memorandum referring to a 'crisis', and 1.1.64 letter to Sir Robert Renwick on takeover threat from Rank.

Unorganised papers: three Unidare folders.

The Simoco papers

These were retrieved from Simoco, Cambridge. Many are in the form of 'Nipper' files, and all are stored in five boxes.

Simoco Box 1: Series of reports prepared by Cooper Bros for Pye's new management in 1966 covering Gibbard, and Pye operations in general. Pye employment contracts to 1940, 1940–1946 main board meeting agendas and Stanley's 1943 report on Pye at war, Pye board minutes 1934–1950 (incomplete). Other files cover 1947 Board of Trade census of production, Cathodeon, patents, W.G. Pye and Cleave Radio. A folder marked 'obsolete documents' has Pye's 16.6.33 agreement with Peter Goldmark on the development of television.

Simoco Box 2: Six files covering various aspects of High-Definition Films. Files from Pye company secretary's office include material on High Vacuum Valve Company (Hi-Vac) 1938–1939, Invicta, and Ether Controls. Personnel files on J.R. Brinkley, F.B. Duncan, J.O. Stanley, E.J.W. (Eddie) Stanley and B.J. Edwards.

Simoco Box 3: Extracts from board meetings 1929–1940 arranged by subject.

Simoco Box 4: British Relay Wireless board meetings 1952–1957, Institute of Directors' board meetings 1952–1956, Associated Broadcasting Development Company (ABDC) minutes 1952–1955, Associated Television (ATV) minutes 1952–1956, eight other ATV files, two files on Television Advisory Committee, including copies of Committee papers 1953–1959.

Simoco Box 5: Three further files on Institute of Directors, one file each on High-Definition Films and ABDC. Two folders of press cuttings.

Minutes

A copy of the Pye main board minutes up to 1965 were generously made available by the Gordon Bussey Collection. The minutes for the crisis period 1965–1966 can be found in COS/TWEM2/2.

Other minutes consulted include Pye Telecommunications and Ekco.

Public Records Office

Documents consulted mainly covered C.O. Stanley's wartime activities.
Military radio: in the AVIA series: 12/184, 22/2312 and 54/189. Also CAB 21/1100 and WO 32/10334.
The proximity fuze: in the AVIA series: 12/184, 22/859, 22/1548 and 53/522.
Radar: in the AVIA series: 7/577, 7/690, 7/1477, 12/160 and 12/161. Also HTT 10/19, 16/36, 140, 701; and T166/80.
Among other PRO documents used were: AVIA 54/1589 (ILS); HO 244/32 and 255/189 (C.O. Stanley and the Post Office); HO 244/558 (Pilkington Committee); and CAB 128/127 (Pye's takeover by Philips).

The Dennis Fuller papers

Two files and two box files include Bell Telephone's *The Transistor* (1961), pamphlets on Pye Walkiephone and Pye radio telephones in industry (1955), undated paper 'Notes on the Magnetic Operation of a Proximity Fuze', notes on apparatus for the investigation of Doppler effects, other Pye papers on the radio proximity fuze and US National Defence Research Committee proximity fuze progress reports.

A Phenomenon of the Thermionic Age

Michael Bell's unpublished life of C.O. Stanley (three bound volumes of typescript) contains much valuable material, not least on Stanley's Irish upbringing and early days in London. Bell was a close friend of John Stanley, and was able to talk to a number of Stanley's close relatives and colleagues who are no longer alive (see Interviews below).

Interviews

The following members of C.O. Stanley's family, friends, colleagues and experts gave interviews:

Professor Jack Allen, John Anderson, Lord Ashburton, Corelli Barnett, Hon Susan Baring, Alan Bednall, Brian Beer, Michael Bell, James Bennett, Ellis Birk, Robert Browning, Brian Callick, C.J. Carter, John Chilvers, Sir John Clark, Derek Cole, Mike Cosgrove, Tony Cowley, Audrey Darkin, Don Delanoy, Richard Eden, Richard Ellis, Sally Emerson, Rethna Flaxman, Jo Fletcher, the late Dennis Fuller, Sir John Gorst, Nuala Hall, Patrick Harris, Peter Hoos, the late Donald Jackson, Ken Kennedy, the late Fred Keys, Sarah Lane, Jim Langford, Norman Leeks, Ian Leighton-Davis, Anthony Lucas, Marjorie McCarthy, the late Gordon Maclagan, Walter Meigh, Dr Patsy Morck, Ted Murphy, Michael Nathan, John O'Connell, the late Bill Pannell, Geoff Peel, Anne Powell, Lord Renwick, Norman Rolfe, Michael Rose, the late Dadie Rylands, David Schoenberg, Tom Scott, Ian Sichel, Julius Silman, David Smith, Nicholas Stanley, David Stewart, Nancy Sturrock, Warren Tayler, the late Peter Threlfall, Willie Wakefield, Daphne Whitmore, John Whitney, Michael Worsley and Donald Zec.

Interviews conducted by Michael Bell

The following gave interviews: S.E. Allchurch, Sam Carn, Norman Collins, Les Davis, Dillon Digby, David Fernie, Hon Roger Frankland, Leslie Germany, Charles Harmer, David Hobson, J.B.H. Jackson, Richard King, Tom Linnane, Bob Piercey, Sir Richard Powell, K.G. Smith, John Stanley, Mrs Bill Tayler (Rue), Reg Thompson and Harry Woolgar.

A–Z files

These contain written communications and some documents: W.G. Allen, John Anderson, Lord Annan, John Beavis, James Bennett (*A Career in Pye*), Joan Bevan, Daphne Bradshaw, John Bullen, R.F. Bullers, Professor J. Coales, Pam Coldham, W.R. Connor, John Cranston, the late Lord Cudlipp (memories of C.O. Stanley), Dr Dillon Digby, Viscount Duncannon, David Dunkley, Mrs D. Dwyer, Richard Ellis, Walter Farrar (military radio), Sir Edward Fennessy (early days of radar), Sir John Gorst (Robert Browning's recollections of Pye; Richard

Meyer's pamphlet on Manx Radio), the late Sir Alec Guinness (working at Arks), Patrick and Liam Harris, Guy Hartcup, Sir Harry Hinsley (Alan Bradshaw at Bletchley Park), Donald Jackson, Damon de Laszlo, Dr E.S. Leedham-Green, Fachtna McCarthy, Louis Meulstee (military radio), Helen Montgomery, Dan Murray, the late Bill Pannell, Dr W.H. Penley, Dr Fritz Philips (papers relating to C.O. Stanley and Philips), David Price, Lord Renwick, Hazel Rose, Tony Sale, Lt Col P.A. Soward (military radio), John Strom, Neil Warnock and Keith Warren.

Books

Among books and other publications the following were particularly useful.

Baldwin, Ralph: 'The Deadly Fuze' (Presidio, San Rafael, CA, 1980).

Barnett, Correlli: 'The Collapse of British Power' (Macmillan, London, 1972).

Barnett, Correlli: 'The Audit of War' (Macmillan, London, 1986).

Barnett, Correlli: 'The Lost Victory' (Macmillan, London, 1995).

Bell, Michael: 'A Phenomenon of the Thermionic Age' (Unpublished).

Benn, Tony: 'Out of the Wilderness. Diaries 1963–7' (Hutchinson, London, 1987).

Bessborough, Earl of: 'Return to the Forest' (Weidenfeld & Nicolson, London, 1962).

Bowen, E. G.: 'Radar Days' (Institute of Physics, London, 1987).

Briggs, Asa: 'The History of Broadcasting in the United Kingdom: Vol. I – The Birth of Broadcasting' (Oxford University Press, London, 1961).

Briggs, Asa: 'The History of Broadcasting in the United Kingdom: Vol. II – The Golden Age of Wireless' (Oxford University Press, London, 1965).

Briggs, Asa: 'The History of Broadcasting in the United Kingdom: Vol. III – The War of Words' (Oxford University Press, London, 1970).

Briggs, Asa: 'The History of Broadcasting in the United Kingdom: Vol. IV – Sound and Vision' (Oxford University Press, Oxford, 1979).

Briggs, Asa: 'The History of Broadcasting in the United Kingdom: Vol. V – Competition' (Oxford University Press, Oxford, 1995).

Bryant, Chris: 'Stafford Cripps' (Hodder and Stoughton, London, 1997).

Buderi, Robert: 'The Invention that Changed the World' (Little, Brown and Company, London, 1997).

Bussey, Gordon: 'Wireless: The Crucial Decade' (Peter Peregrinus, London, 1990).

Bussey, Gordon: 'The Story of Pye Wireless' (Pye Ltd., Cambridge, 1979).

Cairncross, Alec: 'Years of Recovery' (Methuen, London, 1985).

Cairncross, Alec: 'Managing the British Economy in the 1960s' (Macmillan, London/St. Anthony's College, Oxford, 1996).

Clark, Kenneth: 'The Other Half' (John Murray, London, 1977).

Cockroft, Professor J. D.: 'General Account of Army Radar' (Unpublished, 1945).

Cockroft, Professor J. D.: 'Memories of Radar Research', *IEE Proc. A, Phys. Sci. Meas. Instrum. Manage. Educ. Rev. (UK)*, 1986, **132** (6), pp. 327–39.

Collier, Basil: 'The Defence of the United Kingdom' (Her Majesty's Stationery Office, London, 1957).

Dow, J. C. R.: 'The Management of the British Economy 1945–60' (Cambridge University Press, Cambridge, 1965).

Foster, R. F. (Ed.): 'The Oxford History of Ireland' (Oxford University Press, Oxford, 1989).

Geddes, Keith and Bussey, Gordon: 'The Setmakers' (British Radio and Electronic Equipment Manufacturers' Association, London, 1991).

Goldmark, Peter: 'Maverick Inventor' (Saturday Review Press/EP Dutton, New York, 1973).

Guinness, Alec: 'Blessings in Disguise' (Hamish Hamilton, London, 1985).

Hartcup, Guy: 'The Challenge of War' (David & Charles, Newton Abbott, 1970).

Hartcup, Guy: 'The Effect of Science on the Second World War' (Macmillan, London, 2000).

Jones, R. V.: 'Reflections on Intelligence' (Heinemann, London, 1989).

Mayhew, Christopher: 'Time to Explain' (Hutchinson, London, 1984).

Moody, T. W., Martin, F. X. and Byrne, F. J.: 'A New History of Ireland. Vol. III' (Clarendon Press, Oxford, 1982).

Oatley, Sir Charles: 'My Work in Radar 1939–45' (Unpublished).

Owen, Geoffrey: 'From Empire to Europe' (Harper Collins, London, 1999).

Pile, General Sir Frederick: 'Ack-Ack' (Harrap, London, 1949).

Pollard, Sidney: 'The Development of the British Economy 1914–1950' (Edward Arnold, London, 1962).

Pollard, Sidney: 'The Wasting of the British Economy' (Croom Helm, London, 1982).

Postan, M. M. and Moissey, Michael: 'Design and Development of Weapons' (Her Majesty's Stationery Office, London, 1964).

Sampson, Anthony: 'Anatomy of Britain' (Hodder and Stoughton, London, 1962).

Sayer, Brigadier A. P.: 'Army Radar' (War Office, London, 1950).

Savage, Robert J. Jr.: 'Irish Television' (Cork University Press, Cork, 1996).

Sendall, Bernard: 'Independent Television in Britain. Vol. I' (Macmillan, London, 1982).

Sullivan, Mary and McCarthy, Kevin: 'Cappoquin – A Walk Through History' (Privately published).

Watson-Watt, Sir Robert: 'The Evolution of Radiolocation', *Journal of the IEE*, 1946, **93**, Pt I.

Watson-Watt, Sir Robert: 'Three Steps to Victory' (Odhams, London, 1957).

Zimmerman, David: 'Top Secret Exchange' (McGill University Press, Montreal, 1996).

The Stanley Foundation is grateful for help from the following: Col J.A. Aylmer, Piers Brendon, Sir Alec Broers, Alan Kucia, Jenny Mountain, and K.C.A. Smith of the Churchill College Archives Centre. Photographs were kindly provided by the following institutions and individuals: Arks Publicity (Ireland) Ltd; Army Historical Trust; Cambridgeshire Collection; Cavendish Laboratory; Hulton Getty Picture Collection Ltd; Imperial War Museum, Irish National Library, Dublin; National Museum of Photography Film and Television; Popperfoto; the Royal Artillery Library, Woolwich; Royal Signals Museum, Science Museum, the Stanley Foundation, Brian Beer, Gordon Bussey, Dodo Dwyer, Douglas Fisher, Paul Goodman, Nuala Hall, Victoria Hall, Geoffrey Peel, Bob Smallbone and John Trenouth.

Index

356   *Index*